Cognitive Studies in Literature and Performance

Series Editors
Bruce McConachie, Department of Theatre Arts, University of Pittsburgh, Pittsburgh, PA, USA
Blakey Vermeule, Department of English, Stanford University, Stanford, CA, USA

This series offers cognitive approaches to understanding perception, emotions, imagination, meaning-making, and the many other activities that constitute both the production and reception of literary texts and embodied performances.

More information about this series at
http://www.palgrave.com/gp/series/14903

Bruce McConachie

Drama, Politics, and Evolution

Cliodynamics in Play

Bruce McConachie
Department of Theatre Arts
University of Pittsburgh
Pittsburgh, PA, USA

Cognitive Studies in Literature and Performance
ISBN 978-3-030-81376-5   ISBN 978-3-030-81377-2 (eBook)
https://doi.org/10.1007/978-3-030-81377-2

© The Editor(s) (if applicable) and The Author(s), under exclusive license to Springer Nature Switzerland AG 2021
This work is subject to copyright. All rights are solely and exclusively licensed by the Publisher, whether the whole or part of the material is concerned, specifically the rights of translation, reprinting, reuse of illustrations, recitation, broadcasting, reproduction on microfilms or in any other physical way, and transmission or information storage and retrieval, electronic adaptation, computer software, or by similar or dissimilar methodology now known or hereafter developed.
The use of general descriptive names, registered names, trademarks, service marks, etc. in this publication does not imply, even in the absence of a specific statement, that such names are exempt from the relevant protective laws and regulations and therefore free for general use.
The publisher, the authors and the editors are safe to assume that the advice and information in this book are believed to be true and accurate at the date of publication. Neither the publisher nor the authors or the editors give a warranty, expressed or implied, with respect to the material contained herein or for any errors or omissions that may have been made. The publisher remains neutral with regard to jurisdictional claims in published maps and institutional affiliations.

Cover illustration: Anatolyi Deryenko/Alamy Stock Vector

This Palgrave Macmillan imprint is published by the registered company Springer Nature Switzerland AG
The registered company address is: Gewerbestrasse 11, 6330 Cham, Switzerland

Preface: Follow the Science

Drama, Politics, and Evolution began with a simple question that eventually spiralled into new discoveries and previously unknown areas of research for me. There is now substantial scholarship on how polarized and tribalistic Americans have become in matters of politics, religion, and culture. Regarding culture, popular dramatic entertainment in all media is now as polarized as everything else. While there is no doubting this political and cultural gridlock, it's not altogether clear why this is so and how we got here. Perhaps we are polarized primarily for historical-structural reasons. According to this perspective, US economic and political history is mostly to blame for encouraging the two major political parties to seek influence among different groups of the population, solidifying these political alignments, and reinforcing them through the culture wars of the 1970s and beyond. Some political scientists, however, track polarization and the resulting culture wars to what they call inherent bias, the notion that our political nature responds to tribalistic inducements and other predilections that are—at least in part—genetically inherited. Of course these two reasons for the breakdown of social cohesion in the nation are not mutually exclusive. Think how difficult our situation would be if there were both structural and evolutionary reasons for the antagonisms wracking US culture and politics today! Among its other arguments, *Cliodynamics in Play*, my subtitle for the book, will present evidence that this, in fact, is the case.

As we know, in some progressive circles even suggesting that evolution could play a role in our politics and drama is anathema because it supposes that humans are not free to make their own political choices. But how "free" are we? *Homo sapiens* would be an evolutionary anomaly if our species did not have some constraints on our behavior that have been genetically built into our DNA. In evolutionary circles, the name for such an unscientific bias is "blank-slatism," which refers to John Locke's enlightenment notion that human minds, when they are born, are essentially blank slates upon which families, communities, and nations write their instructions as the child is growing up. By privileging what they came to understood as "nurture" over "nature," traditional Lockeans denied that humans had any inherent nature at all. But by the end of the twentieth century, the Lockean's underlying premise for this belief—that evolution does not really matter anymore because behavior is entirely 90% learned—was no longer scientifically credible.

Nonetheless, Locke's legacy had shaped eighteenth hopes for republican governance, exerting substantial influence among Jefferson and other Founding Fathers in the US, for example. Locke's beliefs and ideas may still be seen in the hope expressed by many progressive journalists and academic humanists that if we present the truth simply and logically, we can educate American voters to reason correctly about politics. This faulty assumption reminds me of the subtitle of a 2008 book by cognitive scientist and political consultant George Lakoff. After naming his tome *The Political Mind*, he adds this subtitle: *Why You Can't Understand 21st-Century American Politics with an 18th-Century Brain*. Of course Americans never had "eighteenth-century" brains, per se, as Lakoff knows, but their veneration of the Founders led many citizens to adopt much of Locke's blank slatism. Like Lakoff, I have nothing against truth and logic, but both of us agree with those political scientists who have been telling us for decades that people decide how to vote and also how to judge the votes of others on the basis of many variables. Further, one notable downside to the assumption that voters are rational leads to a recurrent problem among humanists and progressives: blaming the citizens who disagree with them for not "voting their interests." If rationality, however, defined, matters little to most voters, we might better start with an attempt at empathy to figure out what voters' true "interests" might be.

While empathy remains a safely neutral term for most humanists, "evolution" still sets off alarm bells. Cognitive psychology might be an aid to critics and play directors, many will concede, but the popular understanding of evolution—with its racist categorizations and "survival of the fittest" ethic a clear threat to egalitarian justice—remains "beyond the pale" for many academics in the humanities. In retrospect, it's evident that many progressive humanists (including me) were able to join the "cognitive turn" after 2000 because much of the early scholarship in the field segregated cognitive psychology (including understandings of empathy) from evolutionary thinking. While I and others knew somewhere in our frontal lobes that humans only have the cognitive architecture we possess as a result of natural selection, this obvious fact had not seemed very important. Most of the early cognitive work in humanistic fields, including Crane's *Shakespeare's Brain* (2001), Fauconnier and Turner's *The Way We Think* (2002), Bordwell's *Poetics of Cinema* (2008), and my own *Engaging Audiences* (2008), used cognitive science as an aid to analysis and interpretation, without questioning assumptions about cultural relativism and the implications of possible universals in human behavior for the writing of history. Nonetheless, as my own cognitive theoretical and interpretative work developed and I altered my position on evolution, I began to wonder how Darwin and his contemporary allies might be challenging my orientation to history.

Writing *Evolution, Cognition and Performance*, published in 2015, helped me to find out. This book led me to reject social constructivism and the nature–nurture dichotomy for the cognitive–evolutionary paradigm of Enaction. I also incorporated Torbin Grodal's "bioculturalism" from his *Embodied Visions: Evolution, Emotion, Culture, and Film* (2009)—the recognition that biology and culture are fundamentally intertwined—into my 2015 book. Still believing that historical particulars must somehow trump the universals of whatever "human nature" evolution had bestowed upon us, however, I avoided a full embrace of Grodal's "universal" claims. After reading what several evolutionarily oriented political scientists had to say about the political part of our human nature a couple of years later, however, I returned to Grodal with pleasure and understanding; his universalism was now an advantage. According to *Embodied Visions*, the evolutionary foundations of our human nature allow us to use our cognitive and emotional competencies to comprehend and enjoy films and other forms of embodied play. For Grodal, these universals simply work at a more foundational level of our behavior than do cultural norms and other, more historical prompts and constraints.

But Grodal is primarily a theorist-critic and I was also searching for more historiographical grounding than my 2015 effort required. *Drama, Politics, and Evolution: Cliodynamics in Play* charts the outlines of my search and its conclusions. Cliodynamics, the mathematical modeling path I followed as a means of exploring our performative past and its links to significant factors of historical causation, led me to a deeper understanding of our evolutionary roots as a species. As a result of this investigation, I found myself questioning foundational historiographical assumptions, including Raymond Williams's neo-Marxism and several tenets of new historicism, that have long grounded much of my historical writing. I also pushed beyond individual cognition, which can work well to explain the methods and choices of actors, directors, and critics, into new areas of social cognition, a necessary shift for understanding audience groups and framing broadscale historical explanations.

Drama, Politics, and Evolution, then, is much more unconventional than I could have predicted when I started the project four years ago. It crosses boundaries in subject matter (offering insights on religious rituals, stage performances, films, and television), bridges distinct methodologies (quantitative and qualitative modes of explanation and analysis), and attempts to mix epistemologies (claiming universal knowledge in the midst of historical particulars). Although the book has taken longer to write than most of my previous efforts, I find that my inability to end the thing before January of 2021 has ironically been fortunate. Without initially planning for it, I discovered during my final revision that I have a lot to say that is relevant to the present conflicts across the world between authoritarianism and democracy, both of which emerged in key predispositions in our evolutionary past and will continue to play out in the politics and mediated dramatic performances of our future.

Realizing that some readers may want to skip ahead to read the sections of the book that interest them the most first, I have organized *Cliodynamics in Play* in three semi-autonomous parts. This organization also accords with Palgrave Springer's interest in selling chapters and whole parts of the book separately, in downloadable chunks. Although I believe it is best read from front to back, some readers may wish to begin with "Part III, Neoliberal Capitalism and Political Disintegration in the US, 1945–2020," and then jump back to "Part I, Orientations," to check out the theories and methods underlying my later assertions. It's also possible to begin in the middle with "Part II, Political Universals in History and Performance." Part II starts in the deep history of the Pleistocene Epoch roughly two million years ago and traces our evolution from the genus level, *Homo*, to our species, *sapiens*, before chronologically exploring many of the major ramifications of our evolutionary past on human performances, first from rituals to dramas and, later, from Aeschylus to Brecht. After that, the reader could choose between moving back to the theory in Part I that makes such a historical sweep possible or forward to Part III to examine, in finer detail, what evolution has to say about politics and drama in a single nation-state for a mere seventy-five years. While I believe these three parts are indeed semi-autonomous and mostly self-contained, I can't rule out some confusion. So, when perplexed by a name or term, my advice is to try the index and go back in the book to an early reference.

This brings me to the final point of my preface. Writing in the shadow of the COVID-19 pandemic and understanding it as the spear point of climate change for the Earth, I find it difficult to foresee a stable future in which the pursuit of historical knowledge concerning the political effects of dramatic entertainment in all media will be able to be investigated in empirically responsible ways. In the past, historians in theater, film, and communication studies have pursued this and similar kinds of knowledge, but the mounting financial problems of universities around the world will force cuts in budgets that will likely sideline (even further) the research efforts of scholars in these fields. Short term, historians in these areas must forge alliances and cross academic boundaries for mutual support, if similar investigations are to be pursued in the future. If it succeeds at all, *Cliodynamics in Play* demonstrates the many commonalities deriving from our evolutionary roots that have shaped *H. sapiens*' past experience of religious rituals, stage plays, films, YouTube videos, and other exercises in what might be called our species' imaginative performance culture. But also, because there is much more to our human nature than our dramatic and political predilections, I hope this book will invite further investigations into our evolution and history that center on our past performances for each other. Among our natural legacies from the evolutionary past are a keen eye for danger, a willingness to accept strangers who come to our campfires with good ideas, and a recognition of the need to cooperate across tribal lines to achieve the common good. It's time to hang together, as Benjamin Franklin once said, "or most assuredly we shall all hang separately."

Pittsburgh, USA   Bruce McConachie
February 2021

Acknowledgments

Like other scholars, I have benefitted immensely from collegial feedback at symposia, conferences, zoom discussions, and emails during the course of publishing recent essays in anthologies that have contributed to this book. This would not have been possible without the professional organizations that provide the framework for such ongoing scholarship. In particular, I am happy to thank The American Society for Theatre Research, Cognitive Futures in the Arts and Humanities, and the International Network for Cognition, Theater, and Performance.

Regarding individual scholars, *Drama, Politics, and Evolution* is indebted to the insights of Joseph Carroll, Amy Cook, Stan Garner, Thalia R. Goldstein, James Hamilton, Emelie Jonsson, Rick Kemp, John Lutterbie, Lynn Tribble, Peter Turchin, and Donald Wehrs.

Thanks, too, to many at Palgrave who have facilitated the publishing of this book, especially Editor Eileen Srebrenik and Shreenidhi Natarajan, Project Coordinator for *Drama, Politics, and Evolution*.

I am also grateful to Palgrave-Springer Nature for permission to use some of the material in two of my essays previously published in two of their anthologies: Eds. Donald R. Wehrs and Thomas Blake, Bruce McConachie, "The Bio-Cultural Evolution of Language and Prosocial Emotions," *The Palgrave Handbook of Affect Studies and Textual Criticism*, 135–154 (New York: Palgrave Macmillan, 2017) and Eds. Joseph Carroll, Mathias Clasen, and Emilie Jonsson. Bruce McConachie, "Cliodynamics and Dramatic Performances," *Evolutionary Perspectives on*

Imaginative Culture, 255–272 (Cham, Switzerland: Springer Nature, 2020). This material is reproduced with permission of Palgrave Macmillan and Springer Nature. Finally, I have used a few excerpts from my book, *Evolution, Cognition, and Performance*, published in 2015 by Cambridge University Press (Cambridge). These excerpts have been reprinted with permission.

Finally, I dedicate *Drama, Politics, and Evolution: Universals in Play* to my wife, Stephanie, for her patience, wisdom, and proofreading ability. All my love, respect, and gratitude.

Contents

Part I Orientations

1 **Genes, Culture, and History in Coevolution**   3
   1.1 *Coevolution and Human Nature*   3
   1.2 *Sociology, History, and Cliodynamics*   13
   1.3 *Consilience and Cooperative Naturalism*   22
   References   26

2 **Ritual and Dramatic Performatives in History**   29
   2.1 *Evolution, Cognition, and Rituals*   30
   2.2 *Moving Rituals into Thick Explanation for Drama*   36
   2.3 *Empathy, Emotion, and Narrative*   40
   2.4 *Politics, Social Cohesion, and Altruism*   56
   References   59

Part II Political Universals in History and Performance

3 **Coevolution in the Pleistocene**   63
   3.1 *Coevolution in the Early and Middle Pleistocene*   64
   3.2 *Cooperative Cultures in the Late Pleistocene*   70
   3.3 *The Invention of Language*   74
   3.4 *Ritual Interactions in Egalitarian Cultures*   80
   References   89

4 Destructive Creation in World History, 9000 BCE–1700 — 93
- 4.1 From Farming Villages to Macro-States — 94
- 4.2 Mega-Empires and a Democratic Opponent — 99
- 4.3 Accommodating Complexity Through Universal Religions — 109
- 4.4 Bunraku Politics in Tokugawa Japan — 114
- References — 125

5 Capitalism, Tribalism, and Democracy, 1500–1960 — 127
- 5.1 Reformation and Enlightenment in Europe — 128
- 5.2 Mozart in Josephean Vienna — 132
- 5.3 Capitalism and Imperialism, 1500–1900 — 139
- 5.4 European and American Tribalisms, 1770–1945 — 142
- 5.5 Postwar Attempts to Transcend Racist Tribalism — 149
- References — 163

Part III Neoliberal Capitalism and Political Disintegration in the US

6 Postwar Hegemony and the Reagan Reversal, 1945–1985 — 169
- 6.1 The Domestic Consequences of World Hegemony, 1945–1975 — 170
- 6.2 From the Great Society to Reagan's Neoliberalism, 1960–1985 — 176
- 6.3 Dysphoric Films Before and After the Reversal, 1950–1980 — 180
- References — 191

7 Neoliberalism and Political Realignment, 1980–2015 — 193
- 7.1 American Life Under Neoliberalism — 193
- 7.2 Realignment and Polarized Politics — 203
- 7.3 The Better Angels of Our Nature — 209
- 7.4 The Internet and Neoliberalism — 214
- 7.5 Internet News and Politics — 219
- References — 224

8	Neoliberal Politics, Polarized Films, and Authoritarianism, 2000–2020	227
	8.1 Polarization and Authoritarianism	227
	8.2 Polarized Films, 2000–2012	239
	8.3 Rage and Violence in Left- and Right-wing Films, 2012–2018	251
	8.4 Trumpian Carnage in the Time of COVID-19	263
	8.5 Insurrection and Beyond	271
	8.6 The Difficult Path Ahead	279
	References	284

Index 287

About the Author

Bruce McConachie has written books on theater history, historiography, performance studies, cognitive criticism, and the evolution of performance. His major works include: *Interpreting the Theatrical Past: Essays in the Historiography of Performance* (with Thomas Postlewait, 1989), *Melodramatic Formations: American Theatre and Society, 1820–1870* (1992), *American Theater in the Culture of the Cold War* (2003), *Engaging Audiences: A Cognitive Approach to Spectating in the Theatre* (2008), *Evolution, Cognition, and Performance* (2015), three editions of *Theatre Histories: An Introduction* (co-written with others, 2006, 2010, and 2016), and *The Routledge Companion to Theatre, Performance, and Cognitive Science* (with Rick Kemp, 2019). For his scholarly contributions, the American Society for Theatre Research awarded McConachie the Barnard Hewitt and the Distinguished Scholar Awards.

In addition to teaching, directing, and acting at the College of William and Mary and the University of Pittsburgh, McConachie has held visiting academic positions at Northwestern, Warsaw, Helsinki, and Queens (Belfast) universities. Now Emeritus at the University of Pittsburgh, he served as President of ASTR from 2001–2003 and continues to coedit the Cultural Histories of Performance series (with Clare Cochrane) for Bloomsbury Metheun.

PART I

Orientations

As my Preface suggests, my general approach to understanding drama and politics is primarily evolutionary and historical. In the first of two chapters for Part I, I will discuss my major epistemological and methodological choices for the biological and historical framework of *Drama, Politics, and Evolution: Cliodynamics in Play*. Following an initial elaboration of the basics of coevolution—that is, the mutually enabling evolution of genetics and culture for humans—I locate the results of this ongoing process as a part of our universal human nature, with particular attention to our political predilections. This leads me to discuss the impact of the coevolutionary paradigm on the social sciences and their importance in reconceptualizing the dynamics of history. One result of this historiographic shakeup has been the work of biologist and evolutionary historian Peter Turchin. His algorithms lead him to emphasize the importance of warfare and social cohesion for the politically motivated increases in the social complexity of historical polities over time. Turchin's "cliodynamics," as he calls them, will shape many of the political and performative narratives I outline and explain in Parts II and III. Chapter 1 ends with a general discussion of E.O. Wilson's goal of "consilience" among all branches of knowledge in the academy—an epistemological "jumping together" that would allow biological scientists, social scientists, and humanists to proceed from the same assumptions and develop their ideas within the same framework.

Ritual and dramatic performances do not stand apart from Turchin's dynamic systems; as a part of culture, both participate in and cannot escape from its effects. As significant generators of changing symbols, narratives, and norms, performances are broadly "performative," in philosopher J. L. Austin's sense of imaginative cultural actions that are doing something rather than simply describing a situation. Drawing from my previous work on dynamic performative systems in all media, I will use many of those conclusions in Chapter 2 and build upon them in later chapters to come. Some performances in the contemporary world today seem to be reinforcing the disintegrative tendencies of political instability in the world, while others (unfortunately a diminishing few) are apparently helping to shore up one or another of our common cooperative norms. The qualifiers "seem" and "apparently" are necessary here, however, because the present field of film, theater, and performance studies has no widely accepted means of understanding and explaining the interactions among popular performances and the political cultures within which they are flourishing. Like most of the rest of the humanities, the general field of performance history is loaded with "theory," much of it overtly "political," but few theatrical and film historians have tried to test their conclusions through rigorous empirical methods. In Chapter 2, I propose such a method.

CHAPTER 1

Genes, Culture, and History in Coevolution

During the last thirty years, the application of evolution to understanding the past of our own species has produced a significant consensus among many biologists, sociologists, and other social scientists that has implications for how all historians do their work. There is now significant evidence that both genetics and culture operated together during the last two million years on both individual and collective levels to generate the species that eventually evolved to become *Homo sapiens*.

1.1 COEVOLUTION AND HUMAN NATURE

Darwin had advanced the possibility of collective as well as individual evolution, but the notion fell on deaf ears until 1981 when biologists E.O. Wilson and Charles Lumsden published *Genes, Mind, and Culture*. Although other biologists also built models for gene–culture coevolution, Robert Boyd and Peter J. Richerson published *Culture and the Evolutionary Process* in 1985, which offered the first widely agreed upon explanation. Based primarily on mathematical models that showed how genetic changes might have worked with cultural processes to enhance the evolutionary success of our ancestors, it provided substantial theoretical and empirical support for the likelihood of group, as distinct from individual, evolution. The coauthors capped their success with *Not by*

Genes Alone: How Culture Transformed Human Evolution (2005), which directly linked current knowledge about our species' past to the dynamics of coevolution.

The basics of the coevolutionary scenario for our species, which resulted in our ancestors' ability to pass on information through both genes and culture, can be quickly summarized. By roughly two million years ago, the genetic evolution of the *Homo* genus had outfitted our hominin ancestors with physical and behavioral adaptations that enabled them to practice the rudiments of cooperation and culture and to teach them to their children. Early cultures of cooperation began to flourish within some hominin bands and played a significant role in their evolutionary success. Because they could hunt and gather more effectively, cooperative groups survived difficult times to produce more children than other bands, a key to survival under natural selection. Over the course of several hundred thousand more years, the success of the bands that practiced varieties of altruism put pressure on our genome to expand our capabilities for cooperating even more effectively. The result was a dynamic feedback loop between hominin cultures and their genes that substantially improved group dynamics to shape what eventually emerged as our ultrasocial species about 300,000 years ago.

Not by Genes Alone turns the usual either-or, nature-or-nurture question about genes and culture into a both-and truism. "To ask whether behavior is determined by genes or environment does not make sense," say Richerson and Boyd. "*Every* bit of the behavior (or physiology or morphology, for that matter) of every single organism living on the face of the earth results from the interaction of genetic information stored in the developing organism and the properties of its environment" (*ital* in original) (Richerson and Boyd 2005: 266–267). Because a significant part of the hominin "environment" during our evolution was social, it is not surprising that culture would eventually become a coequal factor with genes in the later stages of our evolution.

Boyd and Richerson understand coevolution as a dynamical system and use mathematical modeling to explain how it works. Evolutionary philosopher Kim Sterelny praised the two scientists for deploying dynamical thinking as the "right methodology for investigating the emergence of human social life" (Sterelny quoted in Boyd 2018: 136). Recognizing that "local information" about instances of human evolution would not be sufficient and that evolutionary models of "cooperation" in other animals could not offer plausible explanations, Boyd and Richerson, says

Sterelny, understood that we needed to "advance our understanding through a combination of the comparative analysis of ethnographic and archaeological data; experimental work on individual motivation and cognition; and links between these data streams and formal modeling. These models," concluded Sterelny, "help us to understand how patterns in individual decision and action scale up at the level of populations, and how such populations then change over time" (2018: 136).

The realities of coevolution have sparked significant paradigm shifts in several areas of scholarship that bridge the sciences and the humanities. Many scholars, for instance, have explored the overlaps between human ethics and coevolution for their interrelations. These include: *Mother Nature* (1999) by Sarah Hrdy, *Altruistically Inclined?* (2009) by Alexander Field, *Moral Origins: Social Selection and the Evolution of Virtue, Altruism, and Shame* (2011), by Christopher Boehm, *The Ethical Project* (2011) by Philip Kitcher, and *A Natural History of Morality* (2016) by Michael Tomasello. Like Richerson and Boyd, economists Samuel Bowles and Herbert Gintis use mathematical modeling in *A Cooperative Species: Human Reciprocity and Its Evolution* (2011) to advance the thesis that *Homo economicus*—the prototypical Rational Man pursuing economic gain in the capitalist marketplace still beloved by many economists—has no basis in the deep history of our species. Joseph Henrich's *The Secret of Our Success: How Culture is Driving Human Evolution, Domesticating Our Species, and Making Us Smarter* (2016) emphasizes the key roles played by prosocial emotions and norms during the last two million years of our evolution. Regarding politics, Avi Tuschman wrote *Our Political Nature: The Evolutionary Origins of What Divides Us* in 2013. Also in that year, Richerson teamed up with Morton H. Christiansen, a cultural psychologist, to coedit *Cultural Evolution: Society, Technology, Language, and Religion*, an anthology of 20 essays written by 48 contributors that range broadly among the four subject areas of its subtitle. As the coeditors note in their Introduction, they understand culture as "an inheritance system," constituted by "ideas, skills, attitudes, and norms," that gets passed down through contiguous generations (Richerson and Christiansen 2013: 4). *Cultural Evolution* succeeds in establishing a methodological consensus among empirically based scholars on how to study the coevolution of religion and society, the two areas of the anthology that will be of particular use to my study.

Coevolution is relevant to understanding the roots of political behavior in the history of our species. As will be apparent in Chapter 3, I have synthesized the insights of Henrich, Tuschman, Boehm, and others to discuss how evolution probably produced four clusters of genetically based dynamics that have played out in our political behavior since hunter-gatherer days: **authority**, **homophily**, **equality**, and **altruism**. I say "probably" because there is as yet no consensus as to how coevolution and related processes may have produced what can be loosely termed our political nature. I am putting these sets of politically oriented predilections and the tensions they embody in bold letters when introduced during this chapter and the next one to emphasize their importance to my discussion. I will "embolden" other significant, recurring terms in these chapters.

The primary tension within behaviors centered on authority is between **domination** and **prestige** as the basis for the exercise of legitimate authority in human cultures. Alpha male power, which derived from the genetics of previous great apes on our family tree, dominated the early politics of *Homo* hunter gatherers. Around a million years ago, however, male domination gave way to what evolutionary biologist Henrich and others have called the exercise of prestige, a more cooperative mode of behavior, and a new source of political legitimacy that could also allow for the authority of women and mixed-sex coalitions within hunter-gatherer bands. (Cf., bands are generally smaller and less complex social groups than tribes.) Tensions between the politics of domination and prestige continued for the rest of the **Pleistocene geological Epoch** (2.5 million to 13,000 years ago), changed substantially after our species' transition to agriculture, and remain an important part of human nature and history today.

This political tension was later complicated by **homophily,** our species' inherent desire to separate "us" from "them," which also has deep roots in the behavior of our primate ancestors. Initially a straightforward matter of separating "our" band from "other" hominins nearby, homophily became more complex as band populations increased, divided themselves into subgroups, and made alliances with other bands for marriages and trade. The result was an increasing tension within bands between protecting insiders (later to be termed **tribalism**) and expanding the size of the band to welcome and incorporate outsiders (**xenophilia**). Over time, these differences within homophily led to what we can now see as a range of conflicts between the constrictions of tribalism (in matters of race, religious rituals, national loyalty, etc.) and a xenophilia that could expand to become cosmopolitanism.

The tension within **equality** is between those who practice **egalitarianism** and those who emphasize the importance of natural and/or man-made **hierarchies**. As we will see, our genus likely began the Pleistocene in hierarchically organized bands, like chimpanzees, but gradually moved toward egalitarian behaviors. In general, hierarchical social arrangements align with domination, but there are also those who favor governance by prestige because they wish to elevate in rank those who they believe are naturally more prestigious above others who are not. After the evolution of language, there is little doubt that both ends of these three polarities—domination-prestige, tribalism-xenophilia, and egalitarianism-hierarchy—had their defenders within each band.

As hunter-gatherer life during the Pleistocene Epoch moved away from earlier modes of **domination, tribalism,** and **hierarchy** and toward **prestige, xenophilia**, and **egalitarianism,** a fourth factor, **altruism**, exerted significant influence in the lives of our ancestors, especially through the stories and rituals they were performing for each other around their campfires. Altruism, the genetically based disposition to sacrifice oneself for the survival of another person or group beyond immediate family members, is unique to humans. I will have more to say about altruism, rituals, and dramatic narratives in the next chapter. Hominin hunter gatherers would learn to regulate these four anchors of political life through shared **social norms,** the rules, and standards of behavior expected to be followed by each member of their social group and enforced by the band as a whole. It is likely that the influence of social norms increased as hominin bands increased in size and their levels of **cooperation** grew more complex. As we know, norms for cooperation are particularly important for performances of rituals and dramas in all media of communication.

How can we be sure that the tensions within equality, authority, and homophily and a penchant for altruism and social norms are universal to the behavior of our species? In *Our Political Nature*, Avi Tuschman presents strong empirical evidence that these recurring traits are caused by genetics as well as cultural learning. Because identical twins share one hundred percent of their genetic material with each other, experiments with them are typically the gold standard for testing whether a behavioral trait is partly inherited. In the early 1990s, psychologist Thomas Bouchard tested 88 pairs of identical and 44 pairs of fraternal twins, all of whom were raised in different environments, for their general political orientation on a left–right ("liberal"—"conservative") scale. Numerous previous tests had already established that liberal people in general score more highly than conservatives on personality measurements for openness and extraversion and that conservatives rank above liberals on the trait of conscientiousness. Despite the fact that the identical twins were raised apart and could have learned very different political orientations in their separate familial and local environments, these twins, reports Tuschman, showed "a strong correlation between their political orientations; but the scores of fraternal twins raised separately didn't correlate significantly. These results suggest that genetics plays a decisive role in determining political attitudes" (Tuschman 2013: 26).

Tuschman also presents substantial evidence that the political orientations of citizens in every country across the world can be tracked on a similar continuum. He concludes: "So the left–right political spectrum is universal. It forms a natural, bell-shaped curve [for each national population]" (2013: 29). Most people live near the middle of the curve, different hot-button issues define the extremes at both ends, and various countries have wider or narrower political spectrums. But left and right orientations, as defined by attitudes toward openness, extroversion, and conscientiousness, appear to be universal—another conclusion that suggests a partly genetic rather than a wholly cultural explanation for preferred political values.

Tuschman finds that similar coevolutionary dynamics shape attitudes toward equality and hierarchy, along with an individual's orientation toward domination or prestige as the preferred mode of political leadership, in all societies. Political scientists who have tested groups of liberals and conservatives on their tolerance of political and economic inequality and their belief in the inherent morality of rank and hierarchy have found that conservatives nearly always score higher for these traits than liberals. In addition, individuals expressing these social attitudes usually express the same attitudes toward inequality and hierarchy within their families. Tuschman concludes that "family disciplinary strategies come in two flavors: hierarchical and egalitarian. It's clear that a preference for the former is one of the strongest predictors of political conservatism, while liberals prefer egalitarian morality" (2013: 266). Further experiments reveal that birth order within the family often relates to the political preferences of siblings; early born children will tend to be more conscientious and conservative, while later-borns will typically be more open, adventurous, and liberal. Even though the genetic makeup of siblings is typically very close, how parents raise their children affects their epigenetics— whether the environment "turns on" or "turns off" the expression of specific genetic material—within the family. Evolutionary anthropologists speculate that the primary reason for variance in the political attitudes of siblings probably had to do with the competition for status and food that occurred among them during the long Pleistocene, when birth order and political personality traits could mean survival or death within a hunter-gatherer family. This leads Tuschman to suppose that "the tendency for siblings' politically relevant personality traits to diverge has evolved to increase their chances of surviving the perilous life stage of childhood" (296).

As noted in the Preface, several years before the publication of *Our Political Nature*, cognitive linguist George Lakoff wrote *The Political Mind* (2008), which also emphasized the importance of family relationships for later political orientations. What Tuschman would term "hierarchical" and "egalitarian" family types are very close to Lakoff's "strict father model" and "nurturant parent model" of the family. In his introduction, Lakoff, already widely known as an advocate for the primacy of embodied cognition, discussed three underlying reasons, "none of them obvious," for emphasizing embodiment in politics: "First, what our embodied brains are doing below the level of consciousness affects our morality and our politics…. Second, the forms of unconscious reason used in morality and politics are not arbitrary. We cannot just change our moral and political worldviews at will…. And third, the embodied aspects of mind, as we shall see, connect us to each other and to other living things and to the physical world…. This is how reason really works" (Lakoff 2008: 10–11). Hoping to influence Democratic campaigns in the 2008 election, Lakoff wrote *The Political Mind* to reorient politics away from conventional ways of thinking and toward cognitive realities. Unfortunately, few Democrats took him seriously, which (arguably) helps to explain some of the recurrent ineffectuality of Democratic campaigning today.

Although Lakoff does not take a position on the validity of a universal human nature, his conclusions largely support Tuschman's advocacy of this position. Philosopher Edouard Machery takes up the definitional problems of "human nature" in his essay for *Arguing About Human Nature: Contemporary Debates* (2013), which he coedited with Stephen M. Downes. Machery notes that advances in the evolutionary and cognitive sciences allow us to step away from essentialist conceptions of human nature toward a definition based on coevolutionary realities. He states that this essentialist notion, advanced by Aristotle and others, supposes that "human nature is a set of properties that are separately necessary and jointly sufficient for being a human. Furthermore, the properties that are a part of human nature are typically thought to be distinctive of humans" (Machery 2013: 65). In contrast, Machery proposes that human nature is simply "the set of properties that humans tend to possess as the result of the evolution of their species" (2013: 65). Machery includes such properties as bipedalism, parental investment in child-rearing, and the capacity to use language. On the basis of the discussion of coevolution thus far, we might also include as a part of our human nature predispositions for **cooperation**, **altruism**, and following **cultural norms**, plus Tuschman's more specifically political proclivities concerning **homophily, authority,** and **equality**, which play out in right-left tensions related to **tribalism** and **xenophilia, domination** and **prestige,** and **hierarchy** and **egalitarianism**. Of course these universals, though "properties" common to most humans, are not equally embedded in all of us. Just as some adults will remain disinterested in their children, so others will not care at all about the politics of homophily. Evolution leads to substantial diversity in nearly all species, not to uniformity. "What is required of the properties that are a part of human nature is that they be shared by *most* humans," notes Machery. "Relatedly, the properties that are a part of human nature do not have to be possessed *only* by humans," he adds (*italics* in original; 2013: 66).

Natural universals occur in every area of human behavior, not just politics. Patrick Colm Hogan has investigated and defended the study of literary universals, which includes universals in dramatic literature. Such basic techniques as symbolism, imagery, alliteration, and organizing structures like foreshadowing and plot circularity occur in cultures throughout the world. As Hogan points out, what most humanists actually object to are not universals, per se, but pseudo-universals. He cites Kwame Appiah on "antiuniversalists" who "use the term *universalism* as if it meant *pseudouniversalism*, and the fact is that their complaint is not with universalism at all. What they truly object to – and who would not? – is Eurocentric hegemony *posing* as universalism" (*ital*. in original) (Hogan 2010: 38). Hogan jokes that "no racist ever justified the enslavement of Africans or colonial rule in India as the basis of a claim that whites and nonwhites share universal human properties" (2010: 38). As well, Hogan argues that the study of all universals necessarily draws on investigations of historical and cultural particulars:

> Indeed, it is the opponents of universalism who are most likely to limit our cultural and historical understanding, for the study of universals and the study of cultural and historical particularity are mutually necessary. Like laws of nature, cultural universals are instantiated variously, particularized in specific circumstances. Thus to isolate and test universal patterns, we often require a good deal of cultural and historical knowledge. At the same time, however, in order to gain any understanding of cultural particularity, we necessarily have to presuppose a background of commonality. (40)

Hogan's point is clear: Because universal predilections that work at the level of human nature necessarily depend upon specific cultures to instantiate a given behavior, many historical practices are both universal and particular.

Aware of the resistance of most of his readers to his pitch on behalf of literary universals, Hogan explains the process by which he and his allies in linguistics and literary analysis arrive at the designation of "universal" for a particular attribute of literature. This general method is also used by political and other social scientists to designate as "universal" other types of behaviors in all cultures and societies. "Linguists use the term 'universal'," says Hogan, "to refer to any property or relation that occurs across... languages with greater frequency than would be predicted by chance alone" (2010: 42). There are absolute and statistical universals. "An *absolute* universal is merely a special case – a property or relation that occurs across traditions with a frequency of one. Universals with a frequency below one are referred to as *statistical* universals" (42). Hogan recognizes that the extension of the term "universal" to "statistically unexpected properties may seem odd, even misleading," but "it is perfectly in keeping with standard practices and definitions in all sciences and is inconsistent only with common prejudices about the nature of literary or, more broadly, cultural universals" (42–43). Because "absolute" universals are very rare, my use of the term "universal" throughout this book will refer solely to statistical universals. As we will see, Turchin's cliodynamics, like Hogan and Lakoff's categories, deal in statistical universals. Anthropologists, psychologists, and political scientists are continuing to investigate the political nature of our species today. It is fair to say that there is wide general agreement that universals in politics and drama exist, but also that ongoing experiments and modeling continue to attempt to specify exactly how those universals may be characterized.

Since the election of Trump to the White House, the universal differences between modes of authority based on domination and those based on prestige have attracted particular interest. While some of this research is oriented more generally toward present problems in the psychology of leadership, it is not difficult to understand the results as enduring political problems that have continued from Pleistocene times to the present. Although social-psychological results demonstrate that leadership motivated by prestige and respect is more effective than that based on dominance and fear, some male leaders revert to domination as their chief strategy. Political psychologist Dan P. McAdams characterizes President Trump as a traditional alpha male dominator. Like other alpha males near the conservative end of the spectrum, Trump typically takes all of the credit for the success of his political party and nation, preferring to wrap himself in "hubristic pride" rather than praising others for their work (McAdams 2017: 7). Believing themselves superior by "nature," such dominators also tend to believe that the nature of others is fixed and predictable; they rarely credit others with the ability to grow into a role.

McAdams notes that dominators often share a Hobbesian view of human nature and political action. Because they believe that humans are vicious and life is inevitably a battle, dominators generally fall into one of three political orientations: conservatism, fundamentalism, and authoritarianism. Conservatives emphasize the importance of caution in a dangerous world and often look to traditional institutions for guidance. Fundamentalists find truth and security in past documents and beliefs and often demand adherence to their strictures. "With authoritarianism," says McAdams:

> a dominant authority – typically personified in a single leader – protects the people from danger, takes care of their needs, and consolidates their identification with the group. In order to do so, the dominant leader must make clear distinctions between the 'good' people who are members of the in-group... and the 'bad' people, who are relegated to the out-group....
> (*ital* in original) (2017: 9)

These same general ingredients for authoritarian leadership would have been present when nearby bands of *H. erectus* "others" threatened traditional hunting and gathering grounds during the middle and late centuries of the Pleistocene. But the most successful hominin cultures apparently chose leadership modes of prestige rather than domination.

On balance, leaders with the authority of prestige were probably more effective in generating group cooperation than those who practiced dominance. As Henrich explains, "The reason lies in our cultural nature." Because they are role models for others, prestigious leaders that behave with generosity and altruism "can increase the overall prosociality" of their groups (Henrich 2016: 129).

1.2 Sociology, History, and Cliodynamics

Overall, coevolution has substantial ramifications for the writing of history, including performance history. First, the processes of coevolution continue into the present. Our predispositions for survival and politics that we inherited genetically and culturally from our time in the Pleistocene are still at work today. Substituting "coevolution" for what novelist William Faulkner once said about "history," it is clear that coevolution is "never dead; it's not even past." Second, coevolution is the larger whole of which all of human history is a part. Our evolved predilections, plus the limitations as well as the empowerments of various historical cultures on Earth, have facilitated possibilities and imposed constraints on the paths that human political and cultural history have taken and will travel in the future. This means that history works both through universals—that is, species-wide capabilities for and constraints on human actionas well as through cultural particulars. Put another way, these universal motives for political action find diverse paths for expression in the particulars of every society and culture, imbuing all political performances with a mix of universality and particularity. Indeed, both coevolution and history continue to shape all performances, dramatic and otherwise. As a vital part of culture, performances have always been subject to the larger processes of coevolution. These dynamics will be more apparent in the chapters to come.

Recognizing that a synthesis of the social sciences might provide a basis for joining our species' "deep history" (i.e., from the Pleistocene Epoch to the invention of writing) to our recent, or "shallow" history (everything after that), economist Herbert Gintis has looked closely at the ontological and epistemological claims of anthropology, sociology, political science, and psychology. His *Individuality and Entanglement: The Moral and Material Bases of Social Life* attempts one of the first major syntheses of the social scientific disciplines in order to understand how their insights might work within the dynamics of gene–culture coevolution. Gintis' earlier, coauthored work with Samuel Bowles, *A Cooperative Species: Human Reciprocity and Its Evolution* (2011), had already used dynamics and mathematics to apply the insights of coevolutionary theory to hunter-gatherer economic and political life. In *Individuality and Entanglement*, Gintis acknowledges that although "culture" is a term from anthropology, that discipline cannot be used as the basis for including the other social sciences in a transdisciplinary understanding of culture because all social sciences are not built from the same ontological principles. He approvingly quotes E.O. Wilson's sarcastic complaint that "each discipline of the social sciences rules comfortably within its chosen domain…. so long as it stays largely oblivious of the others" (quoted in Gintis 2017: 267).

Unlike the natural sciences, which build upon the principles of mathematics and physics, notes Gintis, most of the social sciences cannot agree on a common analytical foundation; their epistemologies are not only different, they are incommensurate. Aside from economics and political science, which share the same mathematical basis, sociology, anthropology, and psychology—both as individual disciplines and as related social sciences—says Gintis, "have nothing remotely approaching an analytical foundation." According to Gintis, the situation with anthropology and sociology is even more bizarre: "[T]wo supposedly scientific disciplines studying the same thing – human society – and sharing nothing in common in terms of core theory or empirical data" (2017: 273).

Gintis seeks to overcome the problem of social scientific in-group bias by choosing sociology over cultural anthropology and orienting it toward mathematical model building. In general, Gintis favors sociology as the discipline more relevant to the dynamics of coevolution because, in his synthesis, social processes not only perpetuate our genome, but also facilitate cultural similarities and differences. Society, consequently, becomes the nexus where the coevolutionary forces of genetics and culture meet and interact. Culture remains important in coevolution, however, because it "affects the fitness payoffs to alternative social behaviors, rewarding some and penalizing others. The genes that predispose individuals to behave prosocially, according to the cultural norms of the group, are rewarded by having more copies in the next generation while correspondingly antisocial genes are disciplined by having fewer" (2017: 1). In his reimagining of sociology, Gintis borrows from all of the social sciences to fuse several concepts and methods from classical sociology with current social psychology and sociobiology in order to arrive at an understanding of individual action and social dynamics that can function as both the glue and motor of dynamical coevolution and history. Significantly, Gintis also draws a straight line connecting game theory to *Homo ludens*—man the game player, performance maker, and cultural constructor. Culture in *Individuality and Entanglement* thus becomes communication and symbolic manipulation through games, languages, and performances, at a level of practice that could not occur without the rudiments of social **cooperation** but that also has the ability to shape many elements of social life, including normative behavior.

My periodization for *Drama, Politics, and Evolution* follows the sociological advice of Gintis and draws loosely on the series of "revolutions" laid out by Simon Lewis and Mark Maslin in their *The Human Planet: How We Created the Anthropocene* (2018). Part II of this book tracks major political changes in our species' general history from the Pleistocene Epoch (in Chapter 3) through the agricultural and capitalist revolutions (Chapters 4 and 5), with a focus on how increasing levels of social and political complexity have reconfigured the universal, recurrent tensions evident in our political nature. In contrast, the purpose of Part III of *Universals in Play* is to examine how a variety of historical particulars related to these and other universals have helped to change a single polity over a much shorter period of time as it transitioned from social and political cohesion into gradual disintegration. It begins with a focus on the hegemonic power of the US after WWII and tracks the beginnings of its dissolution after 1965 (Chapter 6), when a neoliberal revolution began to transform American society. Many factors related to declining social cohesion lead the US into steep economic inequality, tribalist rancor, and political polarization, and threaten authoritarian governance by 2020 (Chapters 7 and 8). By necessity, my eight chapters will vary radically in scale and scope. With nearly two and a half million years to encompass, Chapter 3 on the Pleistocene will be organized in mostly 100,000-year chunks. I will shift to 1,000-year, then to 100-year periods of time to discuss significant developments during Chapters 4 and 5. For the 75 years of Part III, I will gear down even further, to decades and half-decades for my three chapters on the US.

To navigate these historical seas, I will turn primarily to the "cliodynamics" of Peter Turchin and others. Like coevolutionary theory and the sociology of Gintis, **cliodynamics,** which fuses the Latin name for the muse of history with a word that conjures structural changes over time, draws on mathematical modeling to investigate problems posed by the past. Regarding the coevolution of cooperation, for instance, Turchin remarks that "if we don't hold ourselves up to the tests of mathematical rigor, it's simply too easy to make logical mistakes and to be led astray by faulty arguments" (Turchin 2016: 82–83). Turchin describes his work in the introduction to his *Ultrasociety* as combining "the insights from such diverse fields as historical macrosociology, economic history, and cultural evolution to build and test models for historical dynamics" (2016: 19). As evolutionary biologist David Sloan Wilson explains, cultural evolution for Turchin "is a multilevel process much like genetic evolution. A learned behavior can spread through a population by benefitting individuals compared to other individuals in the same group or by benefitting the whole group compared to other groups in the vicinity" (Wilson 2019: 182). Convinced that cliodynamics can improve historical explanations, Turchin argues that "traditional history has been deficient" because it cannot empirically distinguish "good explanations from bad ones" (19).

He gives an example testing the consensus among several historians, archaeologists, and political scientists that large-scale, complex societies first evolved in northern China and the Fertile Crescent of the Middle East primarily because of the invention of agriculture. Not so, says Turchin. Agriculture and geography created the necessary conditions for social complexity, he acknowledges, but Turchin drew from coevolutionary theory and masses of data to build a dynamical model that tested whether warfare might be a better explanation than agriculture for increases in social complexity. He suspected as much because the theory of coevolution suggests that wrenching and expensive changes in social complexity are only possible when societies push internal social cooperation as a means to compete against each other, especially in warfare. Turchin also knew that the gradual transition from hunter-gatherer groups to agriculturally based city-states led our species away from the predominance of social **equality** and **prestige**-based politics toward the rise of **hierarchy, tribalism,** and alpha male **domination**. Many of the early emperors, kings, and pharaohs of China, Mesopotamia, and Egypt waged wars of pillage and conquest to elevate their power, enslave their neighbors, and take their land. Turchin's evidence and model building demonstrated that warfare provided a better explanation than agriculture for our species' grudging march toward social complexity. "The costly institutions of complex societies manag[ed] to spread and propagate [between 1,500 BCE and 1,500 CE] because the societies that possess[ed] them destroy[ed] those" that did not (2016: 20), Turchin concluded. The model he built with his colleagues and published in the *Proceedings of the National Academy of Science* in 2013, says Turchin, accurately predicted when, where, and why large empires arose over those 3,000 years.

Specifically, Turchin and his colleagues used what he and others have called "retrospective prediction" to test the two potential explanations (i.e., agriculture or warfare) for advancing social complexity. "For example," says Turchin, "Two rival theories may make different predictions about the behavior of some variable… under certain conditions. We then ask historians to explore the archives, or archaeologists to dig up data, and determine which theory's predictions best fit the data" (Turchin 2008: 35). He adds, "Retrospective prediction, or **'retrodiction,'** is the life-blood of historical disciplines such as astrophysics and evolutionary biology" (Ibid., 35).

As many people around the world have learned recently, predictive algorithms are also the key to epidemiology. Epidemiologists track outbreaks of epidemics to build models of the pace of infectious diseases, test for the primary social and political, as well as epidemiological, variables that appear to have determined their rates of deadliness and spread, then make predictions based on their past data and new algorithms. Unlike cliodynamic historians, of course, epidemiologists cannot begin at the endpoints of their predictions and work backward to explain the reasons for the process; they already know the process and its causes, but must work forwards to explain how it will probably turn out. Both types of model-building based primarily on universal factors have their shortcomings, but both can also be adjusted to take account of new information. Neil Ferguson of the Imperial College in Great Britain, recognized as one of the world's leading epidemiologists, for example, seized headlines in early March of 2020 when he estimated "500,000 deaths in England and 2.2 million deaths in the United States [from the coronavirus pandemic] – if those countries did not take drastic actions" (Wan and Blake 2020: A8). Ferguson's estimates helped to jolt political leaders in the UK and the US into action, but we can no longer know whether his predictions might have been accurate because both nations took some steps to moderate the spread of COVID-19. In this sense, historical retrodiction has an advantage over prediction. Past leaders in history do not get a second chance to moderate their actions; the past is already complete, awaiting the probing questions and dynamical algorithms of the historian.

Nonetheless, Turchin's claims for retrodiction have met with skepticism and occasional rejection from many traditional historians. In "Arise Cliodynamics," Turchin's (2008) argument in *Nature* for transforming history into a science, he acknowledges that most historians distrust mathematics and will resist the search for general causal mechanisms across deep and recent time. Conventional historians, says Turchin, argue that the complexities of historical events and human psychology will always undercut the reliability of retrodiction. "Moreover," he adds, "[they say that] the mechanisms that underlie social dynamics vary with historical period and geographical region. Medieval France clearly differed in significant ways from Roman Gaul, and both were very different to ancient China. It is all too messy, argue the naysayers, for there to be a unifying theory" (2008: 34). Responding with the confidence of an inventor who, when told human beings were not meant to fly, takes you up in his airplane, Turchin points out that "if this argument were correct, there would be no empirical regularities. Any relationships between important variables would be contingent on time, space, and culture" (34). This is manifestly untrue, however, as Turchin has happily demonstrated in studies going back to 2003. And several mainstream historians committed to traditional historiographical methodologies have been convinced. States Ian Morris in a review of *Ultrasociety*, "As his own career shows, there is nothing to stop natural scientists from mastering historical data and applying more rigorous methods to them, with valuable results" (Morris 2017: 154). Empirical rigor is the key to Turchin's success. For example, Turchin follows Richerson and Boyd in demonstrating mathematically how and why a genetic predilection for altruism increased in our species over much of the Pleistocene, creating part of the explanation for the rise in tribalism and warfare after the transition to agriculture.

As noted in the Preface, my interest in deploying Turchin's cliodynamics for *Drama, Politics, and Evolution* departs from my previous historiographical orientation, which drew primarily on the cultural materialism of Raymond Williams. When I first began to combine my cognitive and historical investigations, it seemed to me that cultural materialism was a good fit with the cognitive operations I was discovering and applying to the theater. After all, Williams recognized the importance of audiences as well as the producers of drama in historical cultures and he understood that evolution had prepared our species for cultural life. In *The Long Revolution*, Williams wrote that the "evolution of the human brain and then the particular interpretation carried by particular cultures give us certain rules or models, without which no human being can 'see' in the ordinary sense at all" (Williams 1965: 18) Further, "Particular cultures carry particular versions of reality, which they can be said to create, in the sense that cultures carrying different rules... create their own worlds which their bearers ordinarily experience" (1965: 18). Claiming that her anthology, *Introduction to Cognitive Cultural Studies*, was "compatible with Williams's original vision of cultural studies," historian and critic Lisa Zunshine cited the above passage by Williams in her own "Introduction" to the volume (Zunshine 2010: 5). I agreed with her when the book was published ten years ago and my chapter, entitled "Toward a Cognitive Cultural Hegemony," which celebrated Williams's theory, was among four others in her section on Cognitive Historicism.

There are three major problems with the attempted synthesis of evolution and cultural studies in Zunshine's anthology (and in my essay in the collection), however. The first is Williams's understanding of evolution. In 1965, when he republished *The Long Revolution* from which the passage above is excerpted, Williams followed the general consensus of the time that natural selection had prepared humanity to undertake general cognitive operations, but that evolution did little to specify human cognitive development and learning. Behaviorism remained the dominant paradigm in 1965 for explaining learning and culture—a paradigm that emphasized behavioral conditioning over inheritance and other evolutionary factors. In the passage cited above, Williams notes that evolution gave us brains "and then" particular cultures gave us "rules or models" for interpretating our experiences. This is essentially a behavioral explanation for culture; for Williams, evolution supplied us with basic cognition and culture followed, allowing learned behavior to do the heavy lifting for social practices and beliefs. As I mentioned in the Preface, this model promotes blank slatism, the Lockean belief that cultural learning is much more important for how we live our lives than our evolved human nature. Hogan solves this explanatory difficulty by locating cognitive universals at a more formative level than the learned "rules or models" of particular cultures. The contributors (including me) to Zunshine's section on cognitive historicism, however, do not clarify or resolve the tension between their emphasis on particular cultures and the material, universal realities of coevolution.

Materialism is also at stake in Williams's (and others') versions of cultural studies. Like chemistry and physics, evolutionary biology is rigorously based on material realities—natural environments, cultural remains, and DNA, for example. Until the 1950s, most Marxists around the world could also claim to be following the example of scientific materialism. Their master metaphor for understanding the determining relationship between the material realities of capitalism and the cultures that capitalism produces (social structures, politics, theater, print culture, etc.) was the relation of a base to a superstructure; for orthodox Marxists, the economic base determined the sociocultural superstructure. Although mechanistic and rigidly structuralist, the base/superstructure metaphor emphasized the importance of economics as the prime mover in history and gave hope to Marxist revolutionaries that other changes would surely follow if capitalistic exploitation were overthrown. Along with historian E.P. Thompson, Williams challenged this master metaphor in the 1950s and '60s. His aim, said Williams, "was to emphasize that cultural practices are forms of material production," a goal that undercut the orthodox Marxist divide between economic base and sociocultural superstructure (cited in Palmer 1990: 209). As historian Brian Palmer concludes in *Descent Into Discourse: The Reification of Language and the Writing of Social History*, Thompson and Williams inspired "much positive and exciting [historical] work," but their corrections to the old base-superstructure model allowed their claims to be "too easily incorporated into an emerging orthodoxy... that [shut] its eyes to materialism. The cultural became the material; the ideological became the real" (1990: 210). The result, concludes Palmer, was that when "structuralism gave way to poststructuralism, the way was thus unconsciously prepared for the reification of discourse, even within those very sectors of intellectual life supposedly most committed to historical materialism" (210). By the 1980s, according to Palmer, historical theories that endorsed the ideas of Foucault, Lacan, Derrida, and most poststructuralists had lost their way, drifting toward idealism. But "Critical Theory is no substitute for historical materialism," asserts Palmer; "language is not life" (xiv).

Peter Turchin, biologist-turned-historian, would agree. Although Turchin borrows some insights from Marx, his historiographical practice avoids base-superstructure problems and similar matters of grand theory by considering a wide range of material factors, including economics, and testing them empirically for causal relevance through his models and algorithms. As will be evident in subsequent chapters, these factors include such mundane material realities as state revenues, general population incomes and death rates, the social composition, wealth, and consumption patterns of elite groups, and similar matters that can be reliably enumerated. When paired with the data and calculations provided by Harvey Whitehouse about ritual performances, as we will see, this cliodynamic history can causally connect certain kinds of dramatic performances to historical rises and falls in the social cohesion of a nation-state. By any fair estimation, no contemporary cultural-historical work parading under the banner of critical theory can hope to match the commitment to empirical methods and material realities evident in Turchin's cliodynamic historiography.

In addition, by posing questions that require **retrodiction**, Turchin and Whitehouse only set out historical hypotheses that are demonstrable or falsifiable through cliodynamic calculations. For example, the statement, "Certain kinds of Christian rituals performed in Catholic churches in France for thirty years before the Revolution of 1789 helped to build and maintain social cohesion in the countryside" is a proposition that can be empirically tested. Unlike most historical questions that arise in cultural studies, especially those focused on ideology, this one can be answered by cliometricians with a fair degree of probability. Retrodiction has not been used, however, by historians committed to critical theory. As is well known, Williams relied primarily on ideological orientations to distinguish among dominant, residual, and emergent cultures and historians have used these distinctions to try to understand the role each group of adherents played in specific historical events. While such groupings are suggestive for historical truths, they cannot be retrodictive—predictive in retrospect—without deploying cliodynamic methods across the board. There might be sufficient evidence to classify Catholic rituals in the countryside on the eve of the revolution as evidence of residual culture, for instance, but a historian guided mostly by cultural theory could not empirically claim that such rituals played any role, for or against, in peasant uprisings against rural aristocrats during the "Great Fear" in the summer of 1789. In addition to describing past ideological formations, historians want to know why major events occurred and cliodynamics can provide a more reliable empirical pathway than critical theory to answer such questions.

While cliodynamics has made inroads into the work of some traditional historians, it has had little impact so far on the work of historians and others interested in our Pleistocene past. In their *Deep History: The Architecture of Past and Present* (2011), for example, Andrew Shryock and Daniel Lord Smail, together with several associates in history, anthropology, biology, and archaeology, make a good case for joining the deep evolutionary past of our species to our "shallow" history (i.e., the past that dates from written records), map this (very) *longue durée* in a coherent manner, and present numerous sources of evidence for its further investigation. They do not build models, hypothesize variables, and improve them with **retrodiction**, however. Nor does Yuval Noah Harari in his popular *Sapiens: A Brief History of Humankind* (2015). Like Harari, Patrick Manning's *A History of Humanity: The Evolution of the Human System* (2020) begins the significant history of our species with the invention of syntactical language around 70,000 years ago. This time frame is little more than the blink of an eye over the 2.5 million years of the Pleistocene. It cannot begin to account for the multigenerational operations of coevolutionary dynamics that shaped what Tuschman calls "our political nature." Manning uses occasional modeling and draws on Turchin for part of his work, but his narrow understanding of human political nature cannot encompass coevolutionary complexities.

Although I will draw on some of the insights of *Deep History*, *Sapiens*, and *A History of Humanity* in the chapters to come, Turchin's cliodynamics is the primary backbone for *Drama, Politics, and Evolution*. As we will see, **cooperation** and **altruism**, plus three sets of polar oppositions embedded in **authority**, **equality**, and **homophily**—**domination** vs. **prestige**, **egalitarianism** vs. **hierarchy,** and **tribalism** vs. **xenophilia**—have guided our species' political history since the middle of the Pleistocene. These universal dynamics shift significantly over the *longue durée* of world history and continue to shape the much shorter years of the postwar US.

1.3 Consilience and Cooperative Naturalism

When I began writing this book, I expected I would end with a final chapter discussing the Earth's climate emergency and humanity's difficult political decisions in the coming years of the Anthropocene. Among the probable disasters that await us in the next hundred years, if little is done to reverse present trends, are the release of gigatons of methane gas (CH4) into the atmosphere of the Earth as the permafrost near the North Pole melts, an increase in average temperatures worldwide of 4 degrees Celsius or higher, the mass migration to the North of millions of sick, dehydrated, and malnourished people out of South America, Africa, and southern Asia, the human evacuation of Australia, the spread of the American Desert to the North and East, destroying millions of acres of some of the best farmland on Earth, food riots across the world and the breakdown of governments, the collapse of the West Antarctica ice sheet, leading to the inundation of coastal cities around the globe, and epidemics of cholera, typhus, dengue fever, the Black plague, and several new diseases that will make the 2020–2021 pandemic seem like a summer cold. More than the reported problems of rising temperatures, threatened wildlife, severe storms, and more frequent forest fires, climate change, and resource depletion will result in "cascading" catastrophes, which devastate cultures, destroy nation-states, and wreak havoc on normal social and political life around the globe, according to David Wallace-Wells in his well-researched *The Uninhabitable Earth: Life After Warming* (2019). These effects, of course, have already begun. Some relatively rich people will fare better than the poor, who are already suffering and dying, but everyone born after 2020 who grows to maturity will likely face undreamed horrors. To put these grim realities in the blunt words of climate journalist Roy Scranton, "We're fucked. The only questions are how soon and how badly" (Scranton 2015: 16).

Because questions about "how soon and how badly" still lie in the future and *Cliodynamics in Play* is primarily a history that already promises more than it can easily deliver, I cannot pursue an answer to Scranton's concerns in this book. Addressing this dilemma from a coevolutionary perspective requires a second book, for which *Drama, Politics, and Evolution* will (hopefully) provide a necessary preface. With this goal in mind, I will stay with many of the historiographical choices that I made early on that have shaped my decisions about historical periods, analytical terms, and dramatic interpretations—choices that are particularly relevant to the dynamics of capitalism, neoliberalism, and democracy as well as to matters of social justice in our era of cascading crises. Further, given the relevance of our present pandemic to the fate of humans on a hotter Earth, I see no reason to alter those choices.

This is partly because a coevolutionary approach to history can help us to act responsibly in the dangerous ecology of the Anthropocene. Put another way, anthropocenic thinking is "consilient" with a dynamical approach to history. "Consilience" is E.O. Wilson's term; if two ideas are consilient, says Wilson, they "jump together" (Wilson 1998: 9). I agree with Wilson that we must work toward ontological and epistemological syntheses in the sciences, humanities, and social sciences to ensure that knowledge produced in the academy eventually gains the social and political legitimacy and influence that it merits. "The strongest appeal of consilience is in the prospect of intellectual adventure and, given even modest success, the value of understanding the human condition with a higher degree of certainty," states Wilson (1998: 9). Like Richerson, Gintis, and Turchin, Wilson looks primarily to statistical approaches to provide such validation.

Wilson calls for consilience not only as a way of gaining more legitimacy for empirical methods, but also to improve democratic citizenship and governance: "Most of the issues that vex humanity daily... cannot be solved without integrating knowledge from the natural sciences with that of the social sciences and humanities" (1998: 13). He also believes that democratic citizenship needs to include film, theater, performance, and the other arts. "In many respects," he notes, "the most interesting challenge to consilient explanation is the transit from science to the arts" (210). For Wilson, "the key to the exchange between [them] is not hybridization, not some unpleasantly self-conscious form of scientific art or artistic science, but the reinvigoration of interpretation with the knowledge of science and its proprietary sense of the future" (211). My own and others' push to extend the ideas and some of the procedures of the evolutionary and cognitive sciences into a framework that can provide part of the basis for understanding the political impact of cultural performances is a necessary part of this goal. Nonetheless, there are elements of dramatic history that are not yet amenable to the techniques of modeling practiced by advocates of dynamical systems theory and the math that necessarily accompanies it. While I hope to take Wilson's quest for consilience as far as it can go, the next chapter will make it clear that we do not yet have enough big data on cultural performances in all media to understand how they probably fit into cliodynamic patterns.

Another way to understand the challenge of consilience is through film theorist Murray Smith's four approaches to philosophical naturalism. Most advocates for naturalism, the philosophical position associated with Charles Darwin, John Dewey, and nearly all philosophers of science today, would agree with Peter Godfrey-Smith's general understanding of the term:

Naturalism in philosophy requires that we begin our philosophical investigations from the standpoint provided by our best current scientific picture of human beings and their place in the universe.... The science we rely on is not completely certain, of course, and may eventually change. The questions we try to answer, however, need not be derived from the sciences; our questions will often be rather traditional philosophical questions about the nature of belief, justification, and knowledge. Science is a resource for settling philosophical questions, rather than a replacement for philosophy or the source of philosophy's agenda. (Godfrey-Smith 2003: 154)

With Wilson and others, many philosophical naturalists also believe that science can be a resource for settling questions about evolution, history, and dramatic performances.

Drawing on the "two cultures" debate sparked by C.P. Snow in the 1950s, film theorist Murray Smith distinguishes among four possibilities for separating or joining Snow's "culture of science" from or with the "culture of the arts and humanities." Those in the academy favoring the separation of the two cultures of inquiry could argue for the rigorous exclusion of scientific thinking, a position Smith names "*autonomism.*" What he calls "*freewheeling cherry-picking*" is another anti-scientific position, according to Smith. He calls cherry-picking a "false friend to naturalism, more interested in the magpie theft of isolated scientific discoveries for the purposes of decorating non- or anti-scientific speculations than in combining the methods and insights of the human and natural sciences" (Smith 2017: 3). Unfortunately, such cherry-picking has been rife for two decades now among many who profess to embrace cognitive studies in arts and humanities scholarship, sometimes amounting to what might be called "cherry-pie" books, which taste sweet and go down easily, never troubling non-scientific modes of humanistic consumption.

Among those who favor some kind of naturalistic synthesis joining the epistemologies and methods of the two cultures, Smith distinguishes between "*replacement naturalism*" and "*cooperative naturalism*" (original in itals; Smith 2017: 2–3). Advocates of replacement naturalism argue that the ideas and techniques of the human sciences will one day simply replace traditional humanistic approaches to knowledge. For those committed to cooperative naturalism, which includes Smith himself, "the goal is a genuinely integrative one, where the knowledge and methods of the natural sciences *complement* rather than replace or eliminate those of the human sciences" (2017: 3). I confess to some conflict here. As this chapter demonstrates, my primary orientation so far is closer to "replacement" than "cooperative" naturalism. I have embraced scholars in biology and history who seek, in so far as is possible, to ground their truth claims in mathematical algorithms. While traditional humanists in theory and criticism acknowledge the occasional relevance of such computations in specific circumstances, this approach to knowledge is not widely deployed by most humanists, especially those who focus on the arts. Historians turn to quantifiable reasoning more often, but Turchin, of course, goes beyond most of his tribe in his clear commitment to "replacement" naturalism. Despite Wilson's interest in "interpretation" when it comes to the arts, his book *Consilience* is annoyingly vague on exactly how a scientifically minded humanist might "interpret" the performance of a play or film. I will take up this challenge in the next chapter.

References

Boyd, Robert. 2018. *A Different Kind of Animal: How Culture Transformed Our Species*. Princeton: Princeton University Press.

Gintis, Herbert. 2017. *Individuality and Entanglement: The Moral and Material Bases of Social Life*. Princeton: Princeton University Press.

Godfrey-Smith, Peter. 2003. *Theory and Reality: An Introduction to the Philosophy of Science*. Chicago: University of Chicago Press.

Henrich, Joseph. 2016. *The Secret of Our Success: How Culture is Driving Human Evolution, Domesticating Our Species, and Making Us Smarter*. Princeton: Princeton University Press.

Hogan, Patrick C. 2010. Literary Universals. In *Introduction to Cognitive Cultural Studies*, ed. Lisa Zunshine, 37–60. Baltimore: Johns Hopkins Univ. Press.

Lakoff, George. 2008. *The Political Mind: Why You Can't Understand 21st-Century American Politics with an 18th-Century Brain*. New York: Penguin.

Lewis, Simon, and Mark Maslin. 2018. *The Human Planet: How We Created the Anthropocene*. New Haven, CT: Yale Univ Press.

Machery, Edouard. 2013. A Plea for Human Nature. In *Arguing About Human Nature: Contemporary Debates*, ed. Stephen M. Downes and Edouard Machery, 64–70. New York: Routledge.

McAdams, Dan P. 2017. The Appeal of the Primal Leader: Human Evolution and Donald J. Trump. *Evolutionary Studies in Imaginative Culture* 1 (2): 1–13.

Morris, Ian. 2017. [Review of] Turchin, Peter T Ultrasociety. *Evolutionary Studies in Imaginative Culture* 1 (2): 153–155.

Palmer, Brian. 1990. *Descent into Discourse: The Reification of Language and the Writing of Social History*. Philadelphia: Temple University Press.
Richerson, Peter J., and Morten H. Christiansen. 2013. Introduction. In *Cultural Evolution: Society, Technology, Language, and Religion*, ed. Peter J. Richerson and Morten H. Christiansen. Cambridge, MA: MIT Press.
Richerson, Peter J., and Robert Boyd. 2005. *Not By Genes Alone: How Culture Transformed Human Evolution*. Chicago: University of Chicago Press.
Scranton, Roy. 2015. *Learning to Die in the Anthropocene*. San Francisco: City Lights.
Smith, Murray. 2017. *Film, Art, and the Third Culture: A Naturalized Aesthetics of Film*. Oxford: Oxford University Press.
Turchin, Peter. 2008. Arise Cliodynamics. *Nature* 454: 34–35.
———. 2016. *Ultrasociety: How 10,000 Years of War Made Humans the Greatest Cooperators on Earth*. Chaplin, CT: Beresta Books.
Tuschman, Avi. 2013. *Our Political Nature: The Evolutionary Origins of What Divides Us*. Amherst, New York: Prometheus Books.
Wallace-Wells, David. 2019. *The Uninhabitable Earth: Life After Warming*. New York: Tim Duggan Books.
Wan, W., and A. Blake. 2020. Epidemiologists Face a New Obstacle: Accusations that Their Work is a Hoax. *The Washington Post*, March 28.
Williams, Raymond. 1965. *The Long Revolution*. Harmondsworth: Pelican.
Wilson, David S. 2019. *This View of Life: Completing the Darwinian Revolution*. New York: Pantheon.
Wilson, Edward O. 1998. *Consilience: The Unity of Knowledge*.
Zunshine, Lisa. 2010. *Introduction to Cognitive Cultural Studies*. Baltimore: Johns Hopkins University Press.

CHAPTER 2

Ritual and Dramatic Performatives in History

Turchin and his allies among coevolutionary scholars have made significant progress toward the goal of integrating ritual studies with their rigorous historical methods through the work of Seshat, a global history databank. Founded in 2011 by Turchin and other historians and social scientists, Seshat, according to its website, "brings together the most current and comprehensive body of knowledge about human history in one place. Our unique databank systematically collects what is currently known about the social and political organization of human societies and how civilizations have evolved over time" (Seshat website). (Their site also jokingly explains that the name Seshat's signifies "The Egyptian goddess of databanks.") Seshat now has extensive data on over 400 cultures spanning the last ten thousand years of human life, up to roughly the end of the nineteenth century. In 2017, Seshat examined the social histories of these historical cultures at various stages of social complexity to look for predictable confluences that might demonstrate similar dynamics of development, despite cultural differences. According to the results:

> We were able to capture information on 51 variables reflecting nine characteristics of human societies, such as social scale, economy, features of governance, and information systems. Our analysis revealed that these different characteristics show strong relationships with each other and that a single principal component captures around three-quarters of the

© The Author(s), under exclusive license to Springer Nature
Switzerland AG 2021
B. McConachie, *Drama, Politics, and Evolution*,
Cognitive Studies in Literature and Performance,
https://doi.org/10.1007/978-3-030-81377-2_2

observed variation. Furthermore, we found that different characteristics of social complexity are highly predictable across different world regions. These results suggest that key aspects of social organization are functionally related and do indeed coevolve in predictable ways. (website for Seshat: 2)

For a long time, many anthropologists and historians assumed that historical societies and cultures developed and survived (or collapsed) according to processes that were mostly unique to the historical agents that built and lived within that culture over time. And, in turn, most performance historians assumed that the best way of understanding the politics of the plays and films within such cultures was through ideological critique and historicization. As noted in the previous chapter, the empirical reliability of this relativist assumption has been repeatedly disproven.

2.1 Evolution, Cognition, and Rituals

Included among the directors of Seshat is Harvey Whitehouse, an anthropologist who specializes in ritual studies and religion. Whitehouse, who has made a career of investigating the cognitive and affective bases of religions and the relations among religious rituals, group cohesion, and cultural complexity, supervised the collection and classification of Seshat's data relating to ritual practices at these 400 sites. His 2004 book, *Modes of Religiosity: A Cognitive Theory of Religious Transmission*, summarizes his early work to propose a testable theory of religious transmission that would later be incorporated into the work of Seshat.

Like many other religious scholars, Whitehouse sees religious rituals as dynamic systems stemming from our coevolution that draw upon the default modes of our species for organizing our understanding of the world. "[P]eople everywhere seem to acquire certain similar kinds of information about supernatural agents, rituals, and myths," he states (Whitehouse 2004: 29). Because children at an early age learn the differences between animate agents like themselves and inanimate objects like rocks, it is not difficult for them to conceive of supernatural agents that can inspire other human agents and even move inanimate objects. People also develop naïve notions of empathy to explain why things happen that can easily be transferred to explain the efficacy of rituals. Just as human agents can take actions that cause consequences, so might a god decide to act in a certain way if he or she decides to do so; the usual trick for humans is finding the right ritual to move the god to action.

Telling mythical stories is also universal for our species. Whitehouse draws on Mark Turner's early work in cognitive literary studies to support this assertion. Turner states that "the literary mind is not a separate kind of mind. It is our mind. The literary mind is the fundamental mind.... The mental scope of story is magnified by projection – one story helps us to make sense of another. The projection of one story onto another is a parable, a basic cognitive principle that shows up everywhere…" (Turner 1996: 43). Just as most children are born with the capacity to speak and understand a language, believing in gods, trusting in the efficacy of rituals, and explaining events through projected stories come naturally to our species; our human nature predisposes us to religious belief and **narrative** practice.

Anthropologists Whitehouse and James Laidlaw coedited *Religion, Anthropology, and Cognitive Science* in 2007 to explore ritual and settled on cognitive projection as the key to explaining ritual practices. Building on a synthesis of cognitive and interpretive approaches, Jonathan Lanman and others in the anthology emphasize the importance of what they call the "Hypersensitive Agency Detective Device" (HADD) in the minds of hominins and later *Homo sapiens* as a foundational reason for our species' creation and worship of gods (Lanman 2007: 125–26). HADD specifically draws on our capability for projection. When believers think they can see the face of a deity in the embers of a fire, for example, or are able to perceive the workings of the gods in a thunderstorm, they have projected their own notion of agency into a situation to enable them to understand it as the will of some superhuman power.

HADD builds on the assumption that evolution primed our species to be hypersensitive about invisible agents that might do them harm. Better to interpret that rustling in the bush like a snake about to strike than to ignore the signs of possible agency! On those few occasions when such an interpretation was correct, the hunter-gatherer might survive. Lanman and his fellow authors presume that a belief in unseen higher powers often operated in the same way in our past and probably continues to shape religious belief today. According to anthropologists Lanman, Justin L. Barrett, James Laidlaw, and Harvey Whitehouse, HADD, initially an evolutionary adaptation that sensitized our ancestors to the possibility of dangerous agency nearby, led to the secondary cognitive effect of helping them to invent and perpetuate religious deities with superhuman agency. During the Pleistocene, our hypersensitivity about unseen power facilitated the extension of performance into religious rituals.

Other theorists and investigators have also emphasized our ancestors' search for causal explanations in nature as a primary reason for ritual and religion. In his *Religion Explained: The Evolutionary Origins of Religious Thought* (2001), Pascal Boyer notes that religious representations throughout the world focus predominately on five related themes: agency, predation, mortality, morality, and social exchange. Agency ensures life, predation, and mortality deal with the fear of individual death, and morality and social exchange center on the desire for future life and the success of the band, tribe, or nation. Boyer also points out that most religions tell stories about their central supernatural agents that render them semi-human and believable in terms of causal constraints on their agency. Jesus may walk on water, Mohammed may perform a few miracles, and the Buddha may take some fantastic journeys, but in most respects, these gods eat, sleep, breathe, and walk about the earth just like other mortals. If their superior abilities allowed them to avoid most of the material realities of human life, such as gravity and the need for food and air, it would be difficult for people to understand, value, and remember their superhuman powers. (In this regard, comic book and filmic superheroes are also constrained by some human limitations; even Superman fears Kryptonite. Without such frailties, they would not be as dramatically interesting.)

Scott Atran's study, *In Gods We Trust: The Evolutionary Landscape of Religion* (2002), generally confirms the insights of later cognitive anthropologists who emphasize the crucial role of HADD in the evolution of religious rituals. Atran also notes that religious expression presents what seems to be an evolutionary quandary: Why, he asks, should cultures embrace religious practices that range from stopping work at specific times of the day, to mutilating the body, and (in some instances) to killing beloved relatives simply to keep faith with an immaterial deity? According to Atran, many religious rituals rehearse situations of danger, stress, pain, and occasional death to enable humans to process traumatic experiences in their lives that have no logical or probable explanation or outcome. Such events played a much larger role in human history 50,000 years ago than they do today, when disease, starvation, and violent death were more frequent occurrences. Despite appearances, then, Atran shows that sacrifice, mutilation, and similar ritual practices prepared our species to deal with life-threatening events and to bind individuals more tightly to their group.

As we will see, these same predilections for action, understanding causation, and projecting one story onto others shaped our creation and enjoyment of dramatic fictions. Along with religious rituals, many early dramas also incorporate the same themes and problems enumerated by Boyer and Atran—mortality, traumatic experience, morality, and social exchange. As well, sociologists and anthropologists also understand that ritual practices have helped individuals to bind themselves more closely with fellow members of their religious tradition. In turn, group cohesion has helped our species to mitigate and come to terms with the perduring themes and difficulties that afflict their lives. Whitehouse's 2014 article, "The Ties That Bind Us: Ritual, Fusion, and Identification," written with cognitive anthropologist Jonathan Lanman for *Current Anthropology*, sums up a significant strand of this research. Their abstract for the essay, however, begins by noting two major problems in the scholarship on ritual:

> Most social scientists endorse some version of the claim that participating in collective rituals promotes social cohesion. The systematic testing and evaluation of this claim, however, has been prevented by a lack of precision regarding the nature of both 'ritual' and 'social cohesion' as well as a lack of integration between the theories and findings of the social and evolutionary sciences (Whitehouse and Lanman 2014: 674)

These same difficulties have also bedeviled scholarly progress in integrating evolutionary, anthropological, and historical work on ritual and social cohesion in performance studies. See, for example, the competing definitions of ritual, society, performance, and related terms in Jeffrey C. Alexander, *Performance and Power*, Marvin Carlson, *Performance: A Critical Introduction*, Peter G. Stromberg, *Caught in Play: How Performance Works on You*, and my own *Evolution, Cognition, and Performance*.

Regarding the scholarly evaluation of rituals, Whitehouse and Lanman begin by noting the common observation by many social scientists that all cultures promote two primary means of social cohesion—**identity fusion** and **group identification**. Fusion occurs "when a social identity becomes an essential component of our personal self-concept" (2014: 677), usually leading to a person's strong bonding with a small social group. Group identification avoids such fusion for a looser form of cultural cohesion, which leads people to recognize that they "share certain prototypical features with other group members that are not essential to their individual, personal identities" (677). Identity fusion typically leads to strong **tribalism**, whereas group identification, while not ignoring relevant in-groups, typically facilitates an orientation that is also open to **xenophilia**. The authors add that "social identity researchers have argued that personal and social identities are normally like oil and water – if one is activated, the other is not, and the more one prevails in the individual's social life, the less prominently the other features" (677).

Humans are one of the few animals whose evolutionary predilections allow them to form social relations with others on the basis of both **identity fusion** and **group identification**. Coyotes are another and, significantly, the fusion/identification dichotomy plays out prominently in how they hunt. Fused coyotes hunt in packs, like wolves, whereas coyotes that hunt singly (or in pairs), like lions and other large cats, are expressing simple species identification. Humans, of course, can hunt both ways as well, but when they join an army and primarily "hunt" the enemy in groups, **altruism** kicks in and the "fused" individuals involved will tend to perceive attacks on their group as threats to their genetic kin. Simply "identifying" with the behaviors and norms of a group, however, "will result in intuitions of shared group membership and trustworthiness via the workings of ethnic psychology" (2014: 678). Both practices lead to social cohesion, but while fused individuals may altruistically lay down their lives for their fellow soldiers as for a "band of brothers" in battle, the trust among ethnic allies is more likely to result in bolstering social norms and engaging in mutual aid.

According to Whitehouse, the processes of identity fusion and group identification are each related to a distinctive type of ritual. Rituals that are predominately dysphoric tend to reinforce identity fusion and mostly euphoric rituals usually result in group identification. **Dysphoric rituals** have been called rites of terror; intended to cause disorientation and fear, they often involve coercion, physical punishment, and negative emotions. By causing participants to bond together through the shared memory of a traumatic ritual experience, dysphoric rituals prepare them to participate together in high-risk activities, such as hunting and warfare. **Euphoric rituals**, in contrast, "involve frequently repeated, causally opaque conventional actions with low levels of dysphoric arousal but heavily emphasizing credibility-enhancing displays for beliefs, ideologies, and values" (2014: 681). These rituals persuade primarily through the repetition of a common orthodoxy, rather than demanding obedience to fearful practices. As Whitehouse and Lanman conclude: "Viewed within an evolutionary framework, different societies require higher or lower levels of fusion or identification to fulfill their basic material and economic needs in diverse resource environments... and the two ritual packages evolved, through a process of cultural group selection, to produce the required levels of fusion or identification" (681).

Whitehouse hypothesizes that our species' shift to agriculture led its more complex social groups to gradually abandon dysphoric-fusion rituals for their general population and to feature instead the norm-enhancing values of euphoric-identification rites. A brief 2015 article in *Cliodynamics* written by Whitehouse, Turchin, and a third coauthor, "The Role of Ritual in the Evolution of Social Complexity: Five Predictions and a Drum Roll," predicts how these divergent modes of ritual practice will play out over the centuries when tested by Seshat's algorithms for increasing social complexity. Whitehouse et al. craft five hypotheses, beginning with the statement that after the transition to agriculture "dysphoric rituals correlate with small-scale armed groups, intra-elite conflicts, military revolts, and separatist rebellions" (Whitehouse et al. 2015: 202). No longer necessary for the survival of the whole culture, dysphoric rituals typically fuel the loyalties, resentments, and actions of small groups of non-conformists. (We would expect to find dysphoric predominance today, for example, among the initiation rites of the Proud Boys and similar gangs.) Euphoric rituals of identification, in contrast, perpetuate norms of cooperation needed to sustain increasingly complex social orders.

Seshat's preliminary findings noted in the 2015 *Cliodynamics* piece report that Whitehouse's five predictions were confirmed: "We were finally able to see whether ritual statistically clustered as predicted around imagistic [dysphoric] and doctrinal [euphoric] 'attractor' positions. To our immense relief, they did" (2015: 206). (As the name suggests, an "attractor position" is a major element in dynamic systems analysis. Most systems analysts set up their testing in the hope that these hypothetical "positions" will register a statistically significant rate of attraction.) Back then, Whitehouse admitted, "this was only the beginning" (206). More work by Seshat in the five years since has confirmed Whitehouse's hypotheses linking dysphoric rituals to **altruism** and less complexity and euphoric rituals to group identification and increasing complexity.

2.2 Moving Rituals into Thick Explanation for Drama

What are the implications of these results for later modes of performance beyond religious rituals? Like most of our species' coevolutionary predispositions, our predilections for practicing **dysphoric** and **euphoric rituals** can be expected to continue past the Pleistocene and into the present. Darwin's general rule about descent applies to coevolutionary cultural traits as well as to bones and neurons; descent nearly always involves modification. Just as many of the political practices of hominins during the Pleistocene will not be expressed the same as they were a million years ago, present-day performative practices will likely have morphed into analogous, but still recognizable modes. Consequently, we can expect to find evidence of structural elements and social purposes that remain rooted in the dysphoric and euphoric rituals of the late Pleistocene and early agricultural period in many **dramas**—whether on stage, film, television, or the Internet—performed in more complex societies.

Although rituals are different in several ways from plays and films, most of Whitehouse's categories for coding rituals with his assistants at Seshat can be transferred with relative ease to analyze dramas in complex societies. While some of his categories require a bit of tweaking before they can be applied to dramatic as well as ritual performances, only a few of the less important criteria do not transfer at all. Whitehouse divides the groups involved in rituals into three categories: "specialists," "participants," and "audience." "Priests" and "entertainers" are his primary ritual specialists and his "audience," as might be expected, are those watching the performance but not taking an active part in eating, praying, or similar ritual activities (See Seshat/Methods/Code book/Ritual Variables, [n.d.],[n.p.] for all coded material). These two kinds of roles correspond closely to the conventional parts played by actors-as-characters and spectators in most dramatic productions. Unlike actor/characters, however, religious priests combine self and character into a single, indivisible whole. Whitehouse's "participants," however, are unique to rituals. These may be initiates and highly engaged congregants, but are usually distinct from his specialists and audiences. While I will be attentive to what I take to be the thoughts and emotions of the audience, I will not assume that spectators for dramatic performances (a term I will use interchangeably with audiences) ever became Whitehouse's "ritual participants."

Whitehouse's methodology queries the frequency of ritual interactions among his three groups (specialists, participants, and audiences). In terms of the US entertainment industry in New York and Los Angeles, this question is relevant for spectators, but a different one needs to be asked about actors. Hollywood, for example, has used box office sales, Neilson ratings, and similar data to track the popularity of their products, but looked to Oscars, Emmys, and other awards to rate their performers and the other artists who produced their dramatic shows. Whitehouse's category of "inclusiveness" (whether his ritual specialists and audiences came from the whole polity, the elite, or other groups, etc.), however, is more difficult to transfer to modern modes of producing dramatic entertainments. With few exceptions, most films are cast with professional and star actors, intended for adult audiences, and marketed (in pre-pandemic days) either in prime time or in first-run movie houses. Answering the larger question about the relative inclusiveness of the audience is more difficult. While Hollywood promotes some films for specific audiences, many dramatic entertainment products are intended for wide international distribution. Although European and American theatrical plays are now primarily marketed for upper-income groups, whether film dramas played especially well to elite spectators or appealed primarily to working-class families or minority groups cannot be known without more research. Whitehouse also enquires about the "costs of participation," which—in the case of human ritual sacrifices prevalent in some archaic states Before the Common Era—could be substantial. For most spectators living in modern complex societies, of course, the "costs" have been easier to bear, although some poor families cannot afford online access or the price of occasional movie tickets.

In addition to the social categories above, Whitehouse asks about dysphoric and euphoric elements in each ritual, checks on orthopraxy and orthodoxy (which ensure a standardized performance [orthopraxy] and a standardized interpretation [orthodoxy] of the ritual), and direct evidence in the ritual's content about social cohesion. According to his coding guide, euphoric elements, for example, include such "positive, emotionally arousing components" as feasting, alcohol, drugs, sex, and "synchronous movement," such as dancing and marching. The guidelines for orthopraxy ask about such matters as "the presence of a ritual specialist who polices good practice," which in modern terms translates best to the work of writers, directors, choreographers, union representatives, critics, and similar cultural regulators. For "orthodoxy checks,"

Whitehouse advises his coders to look at the "supervisory prominence of a professional priesthood or other system of governance." Because modern dramatic production typically mixes the supervision of orthopraxy with orthodoxy—aka policing cultural **norms**—some of the same roles (writers, directors, critics, etc.) are involved here as well, often joined by entertainment producers and financial investors. As the ongoing cultural wars in the US and elsewhere have demonstrated, however, arguing about the cultural morality of dramatic entertainment has become an international sport on social media. Finally, regarding explicit ritual content concerning "social cohesion," Whitehouse instructs his coders to look for such evidence as "oath-taking," **tribalism** ("enemies of any group member are my enemy"), and **altruism** ("willingness to die for each other"). Clearly, similar concerns about social cohesion continue to proliferate in all modes of contemporary dramatic entertainment.

Although suggestive and encouraging, the above similarities between the major elements of Whitehouse's code for analyzing rituals and their parallels in producing and regulating modern dramatic performances are not conclusive. The similarities point to evident continuities in the coevolutionary record, but this does not prove that the euphoric plays and dysphoric films of the last hundred years have had the same kinds of effects on the social cohesion of modern polities as their parallel rituals had on their cultures over the last two thousand years. Probably the similarities hold for general causality as well, but as noted in the discussion above, we do not have enough data on audience popularity and response and on the relation of popular entertainment to cultural norms—Whitehouse's orthodoxy checks—to validate the "positive attractors" for modern euphoric and dysphoric elements. The kinds of mathematical tests that Whitehouse and Turchin ran for Seshat to discover dynamic causal relationships joining types of dramatic performances to the social complexity and cohesion of various historical societies cannot be done without extensive data gathering and reliable methods of categorization. Furthermore, Whitehouse's code book simply lists indicators, not explanations. In order to understand the significance of such probable parallels over time, we need to move from surface indicators to causal explanations.

While there's nothing to be done immediately about the lack of reliable data on the dynamics of popular dramatic entertainment, I can begin to explain the details of the cognitive and performative elements that underlie what might be called successful modern rituals of dramatic entertainment. The best approach for such an investigation is probably to explore their major elements in a specific film and/or a theatrical production. I shall do so, with particular attention to social cohesion and the political concerns that abide in our human nature, by turning to Kenneth Branagh's 1989 film of Shakespeare's *Henry V*. While most of my attention will be directed to the film, I will also note some of the major differences and similarities that both separate and unite the theatrical production of the play in 1599 at the Globe Theatre in Elizabethan London with its rendition, nearly four hundred years later, as a film.

Before this can occur, however, I need to introduce Murray Smith's ground rules for understanding films through the lens of scientific naturalism. Specifically, I will deploy much of Smith's method of "thick explanation," his adaptation of Clifford Geertz's methodology of "thick description," and reinforce it with more naturalistic rigor. My goal is a pragmatic position roughly midway between "replacement" and "cooperative" types of naturalism, as previously discussed. Smith emphasizes two scholarly moves that generally distinguish the kind of explanation he has in mind from conventional criticism. The first is to go beyond one's personal response to a performance in order to discuss the major sub- and/or super-personal realities that help to explain its success. While sub-personal realities might involve such factors as neurophysiology and unconscious cognitive operations, the super-personal level should focus our attention beyond individuals to social, political, and historical levels. These factors may include the universal political predilections from our human nature that have been discussed so far. Second, a naturalistic analysis should "strive to prioritize description and explanation" (Smith 2017: 53) and avoid matters of subjective critical judgment. As Smith recognizes, "there is more to aesthetics than aesthetic judgment and more to the study of film than criticism" (2017: 53). In sum, states Smith, "making art, appreciating art, and explaining art do represent three different activities" (54); a naturalistic approach to dramatic art may touch on all three activities, but must emphasize explanation.

Accordingly, I will mostly avoid a critical interpretation of *Henry V* to investigate the underlying and overarching dynamics of the film, with some attention to the play and, before it, to the original production. As in other dramatic films and plays I examine throughout this book, I will focus primarily on the cognitive and coevolutionary elements that help to explain its success with its primary historical audiences. To grasp the political implications of these dramas for the social cohesion of the polity in which these dramas were applauded and appreciated, I will follow Whitehouse's lead and look at the film's dysphoric and euphoric elements, its checks on orthodoxy and orthopraxy, and especially at the content of the film for direct evidence of social cohesion—or its opposite, social disdain, cynicism, and fear. Consequently, I will not be asking questions about artistic intentions, aesthetic choices, and similar matters, unless the answers pertain directly to historical and coevolutionary matters of social cohesion. The other limitation for this discussion is methodological. My primary purpose here is to establish a general approach for applying Whitehouse's questions and procedures to all ritualistic and dramatic performances in the rest of this study. Accordingly, I will summarize the kinds of questions that stage and filmic performances of Henry V raised for audiences and critics in 1599 and 1989, but not delve into the context of both historical events to attempt to answer them.

By the standards of orthopraxy and orthodoxy in 1989, Branagh's *Henry V* was immensely successful. It won high praise from nearly all of the major British and American critics, who generally compared it favorably to Laurence Olivier's adaptation of the drama in 1944, for its gritty, more realistic depiction of London lowlife and the Battle of Agincourt. Like Olivier, Branagh radically cut Shakespeare's script into a useable screenplay (with scenes from both parts of *Henry IV* to include the Falstaff–Harry relationship in brief flashbacks), directed the production, and played the title role. The film was nominated for several Oscars and swept many of the British and European awards, netting Branagh two Best Actors, three Best Directors, and two Best Film prizes in 1990. Using thick explanation to understand its success with predominately English-speaking audiences in 1989–90, I will focus first on the film's deployment of empathy. Brief discussions of the importance of emotions and narrative for my thick explanation will follow.

2.3 Empathy, Emotion, and Narrative

My decision to begin with **empathy** is an easy one. It is clear that the lure of empathy draws spectators toward the agents and actions of all rituals, plays, and films. Worshippers and spectators imaginatively project themselves "into the shoes of" religious and dramatic figures, presented both as symbols and as embodied characters, when they attend performances. In "Dark and Bright Empathy: Phenomenological and Anthropological

Reflections," authors C. Jason Throop and Dan Zahavi reflect on the current state of empathy studies and the term's continuing importance for understanding human sociality, including performance. As their title suggests, Throop and Zahavi combine cognitive and phenomenological perspectives to underline both positive and negative aspects of human empathy. They also insist that "a proper account of empathy has to maintain and preserve its difference from emotional contagion, motor mimicry, sympathy, imaginative perspective-taking, and inferential mind-reading" (Throop and Zahavi 2020: 284). While related to these other cognitive operations and often used in tandem with them, empathy, say the authors, is more evolutionarily foundational: "[E]mpathy is the basic experiential source for our comprehension of foreign [i.e., socially other] subjects and their experiences and... this is what more complex kinds of social cognition rely on and presuppose" (2020: 287). Empathy "is not a theoretical interpretative stepping out of the stream of lived experience to reflect upon another's perspective on the world;" rather, "empathic modes of responsivity... are continuously triggered" in ongoing interactions (287).

Although a direct experience of others, empathy preserves the self-other difference. "What is distinctive about empathy," insist Throop and Zahavi, "is precisely that the empathized experience is located in the other and not in myself" (2020: 289). In this sense, empathy does not encourage the empathizer to fuse with or project themself into the other; it is not an attempt to "identify" with a performer or character. "Empathy is not about me having the same mental state, feeling, sensation, or embodied response as another, but rather about me being experientially acquainted with an experience that is not my own" (2020: 289). While empathizing, the agent uses their own embodiment and experience to learn about the other, but typically such learning is incomplete unless the agent takes the others' emotions, situation, background, and other relevant details into account. How the agent uses such knowledge is also widely variable. The authors ask their readers to "consider how fine-tuned empathic skills – for instance, the ability to detect the disclosures of subtle facial expressions – might aid an interrogator or a torturer whose aim is to inflict psychological harm on somebody" (287). Empathy has a potentially "bright" side as well, but Throop and Zahavi emphasize that confusing empathy with sympathy, pity, or compassion is equally mistaken. Empathizing may lead the agent to a variety of prosocial emotional responses and actions, but that process must not be conflated with such results.

Most social situations in which empathy is in play, of course, quickly move beyond one-way empathizing into dyadic and sometimes multiple social interactions. Although one-way empathy shapes spectatorial response to some theatrical and all mediated dramas today, most rituals and theatrical performances before 1900 flourished in the midst of multiple empathic interactions involving actors and audiences. Throop and Zahavi draw on the phenomenological wisdom of Husserl to point out that "what we find in reciprocal forms of dyadic interaction is the origin of we-acts and thereby of both socialization (being constituted as full-fledged social beings) and communalization (being constituted as a member of a social group and a community)" (Throop and Zahavi 2020: 290). They conclude that "empathy plays such a fundamental role in the fabric of social life, for both good *and* for ill, that its use in ethnographic research is not only significant and generative but also simply unavoidable" (2020: 292).

Philosopher and biologist Evan Thompson also draws on phenomenology to chart the development of the capacity for empathy in human beings. His *Mind in Life: Biology, Phenomenology, and The Sciences of Mind* (2007) helps to underline and specify the foundational significance of empathy for a range of performative interactions. Modifying Throop and Zahavi's injunction to distinguish between motor mimicry and empathy, Thompson locates the beginnings of our capacity for empathy in our "mirror neurons," which activate the mother-infant bond. Networks of neurons in our brains effectively "mirror" or simulate the intentional motor activity produced by another person and perceived by the empathizer. If a spectator watches a performer engage in sword play during the action of a play or film, for example, the same group of neurons in the empathizer's brain is activated as in the player's brain; neurologically, it is as if the observer were engaged in a sword fight herself. By working through our perceptions, bodies, and minds, our networks of mirror neurons unconsciously simulate the actions of others. In this sense, mirror neurons engage in a kind of physiological version of mimesis, the imitation of another's action that Aristotle and many others have understood as the basis of drama. According to cognitive psychologist Giovanna Colombetti, "[W]e have ample evidence that [covertly] mimicking others facilitates interaction, increases liking among participants, and promotes prosocial behavior more generally" (Colombetti 2014: 194). Although empathy begins for humans in mother–infant bonding, we also know that this impulse may later be turned against social others.

Thompson calls the activation of mirror neurons "sensorimotor coupling" and notes that subsequent levels of empathy are more reliant on higher order consciousness than the one that precedes it. The second, more complex form of empathy, which he terms "imaginary transposition" (Thompson 2007: 395), builds upon sensorimotor coupling to engage perspective-taking. The ability to understand how another person perceives the world often combines emotion and perception toward this end. In order to link sensorimotor coupling to imaginary transposition, Thompson draws on recent evidence about child development. By about nine months, the normal child can use other cognitive skills, plus the knowledge and memory that they have gained from mirroring, to engage in imaginary transposition. This begins with the child's recognition that other humans are intentional agents "like me" and leads the developing baby to project her/himself into the mind of a nearby adult.

Soon after children can accomplish the basics of imaginary transposition, they gain the ability to play with "reiterated empathy." Thompson borrows this term from phenomenologist Edith Stein to denote the ability to conceptualize oneself from the point of view of another person. As he explains, "Empathy thus becomes reiterated, so that I can empathetically imagine your empathetic experience of me and you empathetically imagine my empathetic experience of you" (Thompson 2007: 398). Although the cognitive side of reiterated empathy is more complicated and more conscious, it continues to engage our emotions. For toddlers, this ability often signals the onset of shyness and/or confidence, because the child can now begin to perceive how others feel about them. In adults, reiterated empathy may register as embarrassment, indifference, or even guilt, depending on how successfully a person believes others have interpreted their intentions and emotions.

Thompson notes that the fourth level of empathy, "moral perception," is "not the same as any particular feeling of concern for another, such as sympathy, love, or compassion. Rather, it is the underlying capacity to have such other-directed and other-regarding feelings of concern" (Thompson 2007: 401). Reiterated empathy rises to moral perception in children when they begin to understand others and themselves as agents who can take action in the world. Among Thompson's four levels, it is the most conscious and intentional. Nonetheless, despite its later links to cultural **norms** of right and wrong, moral perception is initially rooted in the experience and imagination of the self, typically when five-to-seven-year olds attempt to understand the feelings, plans, hopes, and goals of others. Like the step before it, the fourth level incorporates others as it builds upon and incorporates emotions and perspective-taking. While this empathic level, like the other three, is a general part of human nature, it is also clear that some children who grow into adults do not reach it.

All four levels of **empathy** are in play in Branagh's first scene of *Henry V* (which—following standard citation practice for Shakespeare's plays—I shall designate as I,i). Thompson's first level of empathy is an everyday operation that kicks in unconsciously whenever we are engaged in or observing social interactions. When actors playing the Bishop of Ely and the Archbishop of Canterbury enter and begin to collude with each other to avoid the King's confiscation of some church lands by convincing him to invade France, spectator mirror neurons will pick up on their facial expressions and physical interactions and imaginary transposition will help them to interpret the characters' sneaky intentions. After the entrance of the King accompanied by his nobles, Canterbury presents a mumbo-jumbo argument about Salic Law as justification for the invasion and the churchmen carefully watch King Henry; they deploy reiterated empathy to try to discover if the King's empathic understanding of their actions might cut through their selfish motives and give them away. When the King inquires if Salic Law is sufficient justification, Canterbury reminds him of the past success in France of his famous great-uncle, Edward the Black Prince, two noblemen flatter the king's power and bravery to argue for invasion, and Canterbury chimes in again with a promise of financial aid from the church. The young King, however, makes no more inquiries about the justification or feasibility of war with France. Later in the scene, when the French Ambassador presents the King with tennis balls from the Dauphin "as meeter to your spirit" (I,1, 255) than fighting battles, King Henry ignores this insult to his honor and bids his noblemen make ready for France. But the fourth level of empathy, moral perception, may have sparked an important question in the mind of some spectators: Does Branagh/Harry have any idea what he will be up against when he gets to France? More importantly from an ethical perspective, should he be risking his reputation for honor and upstanding morality, plus the lives of countless soldiers, in a war that two churchmen of dubious loyalty cooked up to avoid losing some real estate?

Of course Shakespeare planted this doubt and Branagh underlined it in his approach to the first scene to awaken their spectators to the ethics of kingship in wartime and to use their powers of empathy to eventually enjoy and take pride in the King's subsequent growth in moral understanding. How this happens is chiefly a matter of manipulating appropriate emotional responses and the crafty handling of narrative structure. By the time Shakespeare's St. Crispian's Day speech arrives for the climactic scene of the film, Branagh's King has been tested in action for honor, honesty, kingly morality, and even piety before God.

The final burnishing of the King's ethical gloss occurs in the night scene before the battle, when he travels incognito among his troops to joke with them, steady their nerves, and provide what the Chorus aptly calls, "A little touch of Harry in the night" (IV, Cho, 47). Near the end of this long scene, after several successful encounters, Branagh/King meets a common soldier, Williams, who, worried about his likely death, is "afeard there are few die well that die in battle" (IV, i, 134). Williams conjures a vision when "all those legs, and arms, and heads, chopped off in a battle, shall join together at the latter day" (IV, i, 128–30) and cry to God for justice against the King who led them in an unjust cause. Harry argues—correctly, according to medieval Christian theology—that each man is responsible before God for his own actions, and besides, the King's war is just. But the image of hacked body parts crying for revenge colors the rest of their scene together. Williams picks a quarrel with Harry over whether the King might avoid death in battle by giving himself up for ransom, a common practice among the nobility and a cause for cynicism among common soldiers. Branagh, in disguise, not only denies that the King would ever do such a thing, he even exchanges "gages"—armored gloves, symbolizing a challenge to a duel—with the commoner and agrees to meet him after the battle to settle their differences in combat. While this affront to feudal **norms** of hierarchy was certainly more scandalous to spectators in 1599 than 1989, Branagh's treatment of it still registers as a surprising act of kingly grace and control. A short time later, Branagh/Harry, soliloquizing in prayer, pledges to God to do more good deeds in expiation for his father's mortal sin of executing King Richard II and seizing his throne, which Harry has now inherited. In short, by the morning of St. Crispian's Day and the battle, all spectator empathy, which likely resulted in suspicion at the start of the film, has been transformed into affirmation and sympathy for the King, just when "the poor condemnèd English" (IV, Cho, 22) need it most.

If empathy has been difficult for cognitive scientists to define, **"emotion"** has tied them in knots. Among the current frameworks for understanding human cognitive and affective operations, enactivism offers one of the most promising approaches for fully integrating their relationship. Thompson, one of the initiators of enactivism, adopts this approach as the framework for his understanding of empathy and I have relied upon it to structure many of my insights in *Evolution, Cognition, and Performance*. In brief, enaction scientists and philosophers reject a computer-based understanding of the brain and body to build upon Varela, Thompson, and Rosch's 1991 book, *The Embodied Mind*. The empirical science behind the enactive paradigm demonstrates that all animal life is constituted by five closely intertwined processes: autonomy, sense-making, emergence, embodiment, and experience. The first of these, autonomy, provides the foundation for the other four. Also known as autopoiesis, autonomy refers to the ability of all animate systems to constitute, sustain, and protect their biological identities. This mode of action involves the ability of all animals from amoebas to *Homo sapiens* to deploy whatever cognitive and emotional abilities they possess in order to survive and flourish. To do this, animals engage the other four processes of action: they make sense of the social and physical environments that constrain and enable their interactions, pursue the meanings that emerge for them from these encounters, draw upon their bodies as well as their brains to understand such experiences, and use both to transform their embodied identities over time.

Colombetti's (2014) monograph, *The Feeling Body: Affective Science Meets the Enactive Mind*, is one of the most successful recent attempts to understand human emotions from an enactive perspective. A definition of emotion compatible with the enaction paradigm must link emotions to perceived meanings. Copious empirical evidence supports the finding that individuals use their emotions to perceive the world as meaningful, not simply as a neutral vista awaiting attribution. In contrast, several other approaches to emotion assume that our appraisals of the environment within and around us occur separately, distinct from our embodied experiences of affect and emotion. Colombetti's work demonstrates that cognitive processes, such as appraisal, cannot be separated from emotional response; both work together throughout the mind and body.

Colombetti also emphasizes the action-based and durational nature of our emotional-cognitive lives. She shares with most scientists the insight that our emotions exert pressure on our behavior; emotional responses carry with them specific action tendencies. Further, emotional episodes tend to last for a while; their durations are "an inescapable, pervasive dimension of brain activity on which sensory information impinges and from which action progresses" (Colombetti 2014: 64). In addition to triggering action, our emotions also set the stage for meaning-making. The same emotion may shape several iterations of a person's perception–action cycle, the moment to moment dynamic system that ensures autopoiesis, before that emotion subsides. A woman or man undergoing an episode of anger, for instance, will likely perceive something in their environment that heightens it and then take action that continues and justifies the anger, repeating the cycle until a new emotion emerges and the previous episode is over.

Anger, of course, is not a prosocial emotion. Prosocial emotions, by definition, tend to generate actions that lead to sociality, happiness, **cooperation**, and **social cohesion** among groups of people. Prosocial emotions may even lead group members to acts of **altruism**, the willing sacrifice of the self for the group. Many prosocial emotions, such as sympathy, admiration, play, gratitude, and loyalty, usually lead to immediate positive results. Others may not generate good feelings right away, but they tend to work toward social cohesion in the end. When a person experiences the emotion of public shame (or even its less extreme cousin, embarrassment) that person generally understands that a heart-felt apology followed by some good deeds is the best way out of a shameful situation. Similarly, working through personal guilt—the psychological realization that you have not lived up to your own ethical expectations— is often best treated by changing your life so as not to hurt those who suffered from your past actions. If the immediate experience and/or long-term results of a specific emotional episode tend to lead to group happiness and solidarity, that emotion may be termed prosocial.

Prosocial emotions are not the same as what Antonio Damasio and some other psychologists and neuroscientists term "social emotions." For Damasio, the social emotions are a "secondary" class of emotions, less prominent and important in human interactions than those emotions that deal directly with survival, such as fear, lust, and disgust (Damasio 2003: 44–49). This distinction is already confusing and improbable, however; it assumes both that our more basic emotions do not play out in significantly "social" ways and that sociality itself has little to do with survival. Further, Damasio's social emotions are not necessarily prosocial. His secondary class includes sympathy, shame, gratitude, and most of the other prosocial emotions mentioned earlier, but also adds humiliation, indignation, jealousy, and other emotions that rarely lead to increased social happiness and cohesion.

For her part, Colombetti dismisses Damasio's (and others') distinction between primary and secondary or social emotions. This distinction, she says, "is based on an increasingly controversial view of the brain (although arguably still an influential one) according to which emotion and cognition are neurologically distinct, with emotion residing in the deeper areas of the brain and cognition depending primarily on the higher areas" (Colombetti 2014: 42). Although it is likely that many mammalian emotions evolved earlier than higher order cognitive skills, the modern brains of *Homo sapiens* have fully integrated emotional and cognitive processes and distributed both over the entire brain. What Colombetti and others identify as our prosocial emotions likely shaped hominin behavior in significant ways during the Pleistocene Epoch.

While this discussion has emphasized positive, prosocial emotions, negative emotions ranging from annoyance and sadness to fear, rage, and disgust are just as important, often more so, in dramatic interactions and social-political circumstances. Sometimes negative emotions emerge as direct feelings in the body; Colombetti notes people reporting a faster heart rate when they experience fear or the contraction of the stomach or throat when feeling disgust. In many instances of heightened positive as well as negative emotion, "the body may be highly present, even when one does not pay attention to it and is rather immersed in the situation" (Colombetti 2014: 122). This applies to dramatic as well as social situations. Theatrical comedies have long trafficked in comic stereotypes that have painted whole classes as figures to be laughed at for their stupidity, arrogance, and/or grotesquery. Such negative responses to individuals, groups, and classes judged to be ridiculous, vain, selfish, angry, and/or vile can turn audiences against them, with consequences that fragment **social cohesion** and lead to destructive varieties of **tribalism** and **domination**.

Following Shakespeare, Branagh's *Henry V* paints the lowlife characters of Eastcheap with the broad brush of comic caricature. Led by Pistol, a braggart-warrior type whose quick anger continually defeats him, the drama's male trio of grotesques also includes Bardolph, whose one claim to former fame—his flaming red nose—has now been extinguished, and Nym, a sullen wretch who can barely find words for his anger against Pistol for stealing his woman, Hostess Quickly, and marrying her out from under him. The effect of their squabbling in II, i of the film is as much pathetic as amusing. Shakespeare likely intended a similar response to highlight how the loss of Falstaff,'s presence in the tavern scenes—the center of comic enjoyment in both parts of *Henry IV*, the previous plays about Prince Harry's rise to fame and throne—has changed the lives of the tosspot-poor of London. Knowing that many Globe theater-goers in 1599 would have recalled Falstaff and Prince Harry's delightful and fateful exchanges and moved those memories into their experience of *Henry V*, Branagh as screenwriter decided to use brief extracts from those two plays, principally *Part I*, as flashback scenes in his film. The appearance of a remembered Falstaff as a kind of Father Christmas figure contrasts sharply with the pettiness, disdain, and anger of the present grotesques in the film. In addition to setting up a warm reception for Judi Dench's monologue as Mistress Quickly about Falstaff's death in II, iii of the play, the flashbacks also allow modern audiences more insight into Harry's checkered past and the debt he now owes to England as a monarch. On a more personal level, the brief scenes also encourage the audience to sympathize with Harry for turning away from an old friend, however debauched and cynical, to take up the duties of kingship.

When the tavern grotesques arrive in France as a part of the invading army, they are quickly revealed as cowards and cutpurses by the regular soldiers who accompany the King. Led by Welsh Captain Fluellen, this loose group of able soldiers from the hinterlands of the British Isles is made up of Irishmen, Scotsmen, and others who, presumably, have declared their loyalty to the English King despite differences in their ethnic backgrounds. After Harry urges his soldiers, "Once more unto the breach.... Cry 'God for Harry! England and Saint George!'" (III, i, 1–34) and leads his soldiers into embattled Harfleur, Bardolph hangs back at the rear with Pistol and Nym, shouting "On, on, on, on, on! To the breach, to the breach! (III, ii, 1). Fluellen finds the three laggards and upbraids them for their cowardice as they slink away. In a later scene, the audience learns that Bardolph might be hanged for stealing from a church and Pistol appeals for help from Fluellen, who refuses to intercede for him because the King has ordered his soldiers to refrain from looting. Morally, *Henry V* sides with feudal order over past friendships and ethnic loyalty when the King recognizes regretfully that Bardolph must be hanged and gives the order to do so. Obey orders in wartime, whatever your ethnic background, or suffer the consequences is the implicit lesson here. Like the play, the film embraces some rival interethnic comedy among Fluellen and his friends, but moves beyond it and beyond narrow **tribalism** to highlight feudal justice by showing the wet and ragged army passing under Bardolph's hanging body on its way to Agincourt.

The other significant group of characters in *Henry V* is the French nobility. For the most part, they are characterized as self-involved, backbiting, and arrogant, especially the Dauphin, the evident parallel to Harry on the French side in age and royal status, who is also portrayed as stubborn and cowardly. Likely knowing that he could not risk his spectators' dismissing the French as unworthy opponents without diminishing the miracle of Agincourt, Shakespeare gave them a King who, though initially slow to act, rouses his lazy noblemen with his knowledge of English military prowess and directs them to answer the present threat. As played by the eminent actor Paul Scofield, the old French King combines fatherly concern and weary authority. The other Frenchman who emerges as an honorable adversary is Mountjoy, a herald who shows respect for Harry, even while delivering disdainful messages from his masters in the French army. King Harry learns to appreciate the herald's candor and even trusts his report of the French and English dead after the battle.

In addition to empathy and emotion, the other fundamental determinant of structure and meaning for performances is story. **Narratives** work for audiences primarily through an effective arrangement of their major embodied simulations or, roughly, what Aristotle called their dramatic actions. Grodal provided a solid definition of "story" in 2009 that still stands:

> A story is a sequence of events brought into focus by one or a few living beings; the events are based on simulations of experiences in which there is a constant interaction of perceptions, emotions, cognitions, and actions. An example: Harry sees the dragoncoming. He is upset, thinks he needs to grasp his sword; he does so and he kills the dragon.... We experience stories as representations of exterior worlds and they may be described as such, but at the same time they represent internal mental and physical processes that have to follow the innate specifications of the body and brain. (Grodal2009: 159)

Like empathy and emotion, many of the "specifications" that shape narratives come to humans courtesy of our species' coevolution.

Although these specifics are not yet as well understood as mirror neurons and the synapses that shape our emotional responses, it is likely that the cognitive pattern for narrative follows what philosopher

Mark Johnson first called the "source-path-goal" universal image schema (Johnson 1987: 113–17). Cultures around the world tell a variety of stories, but nearly all of them begin with a "source," or causal starting point, follow a "path," a sequence of contiguous locations, and end in a "goal" with the resolution (or not) of the initial problem. Grodal's Harry and the dragon story above—strikingly similar in outline to Shakespeare's story about Henry V, of course—is a prototypical example. It may be that the source-path-goal image schema emerged early in the Pleistocene, with hunting parties of men and gathering groups of women going out from their camps (source), seeking game and edible plants (path) for their hominin band, and successfully returning to share it with their families (goal). Early attempts to communicate and repeat their narratives to others around the campfire may have solidified the story pattern.

Though usually classified as a "history" play, a better generic tag for Shakespeare's *Henry V* would be "action-adventure" story. Like Homer's *Odyssey*, *Tom Jones*, *Uncle Tom's Cabin*, and numerous other epics and novels, action-adventure stories often pack in physical motion, social conflicts, heroes, and quests, all of which may end successfully or sadly. Grodal wrote in 2009 that *The Last of the Mohicans*, *Titanic*, *Butch Cassidy and the Sundance Kid*, and *Saving Private Ryan* are good examples of recent action-adventure films from Hollywood. He added that war stories may be understood as a subgenre of the general type. Like most narratives about war, *Henry V* builds up to and finally focuses upon a climactic battle that allows the hero to fulfill his (or her; we can certainly include Marvel Comic's 2017 *Wonder Woman* as a more recent example) quest.

Many war stories, *Henry V* among them, center on important episodes of **altruism**. David Sloan Wilson explains in *Does Altruism Exist? Culture, Genes, and the Welfare of Others* (2015) that altruism is closely related to the group dynamics of our evolution. I will briefly discuss those dynamics in Darwinian terms before turning to an analysis of *Henry V*. As previously noted, when one individual benefits another at an immediate cost to themself, that person has performed an altruistic act. When such individuals join groups to work together for some common purpose, they knowingly delay their own gratification to work for the success of the whole. In his book, *Cognition in the Wild,* evolutionary philosopher Edwin Hutchins used the example of a ship's crew to explain the evolutionary advantage of altruism for teamwork. In this instance, our penchant for altruism activates all crew members to sacrifice some individual freedom and control to enable all to work together so that the ship may sail. Other primate species can cooperate to a degree, but only *Homo sapiens* are willing to sacrifice short-term individual interests beyond their immediate families for long-term group gains. In considering how altruism probably evolved among our ancestors, David Sloan Wilson and E.O. Wilson concluded in a 2007 article that at some point during the Pleistocene, hunter-gather bands figured out that they could more successfully compete against other bands nearby if they cooperated with each other instead of pursuing their own individual plans. Without an army primed for cooperation, Shakespeare's Henry V could not have turned his troops into a "band of brothers."

The key to the psychology of **altruism** for warriors is the hormone oxytocin, which bonds members of small groups pursuing the same goals to each other. Bonding frequently engages empathy among soldiers, as they seek to understand each others' situations and feelings. But oxytocin also powers an aversion to outsiders and strangers; in battle, these "others" easily become "the enemy." This elevation of the emotions that drive **tribalism** is as true for dramas about battles as it is for the real thing. Dramatists, actors, and spectators are primed by altruism and empathy to turn dramatic battles into "us" against "them" conflicts. Of course empathy also has a fourth level, moral perception, which playwrights and directors can deploy to point the way toward peaceful resolution when the fighting is over.

Action-adventure novels often feature third-person narration and Shakespeare deploys a similar point of view by framing the story of *Henry V* with a Chorus, a performer who introduces each act of the play and establishes what Patrick Colm Hogan calls its "storyworld"—that is, its setting in time and place, its primary characters, and often its general mood at the start of new episodes (Hogan 2018: 133). Like Shakespeare, Branagh keeps Derek Jacobi's Chorus—dressed as a modern-day commentator who possesses omniscient vision and shares his concerns about developments in the story directly with the camera—on the edge of the action, providing continuity into the changing storyworld of the show. Although the device of the Chorus potentially allowed Branagh to move toward a more playful, less realistic style for his film, Branagh decided to stay close to the usual conventions of Hollywood realism for action-adventure narratives. This choice likely set up his audience to expect that its storyworld, despite its Chorus and antiquated language, would fulfill many of the desires prompted by other action-adventures films.

In fact, it might be argued that Branagh's film does a better job of fulfilling action-adventure expectations than a theatrical version of the play. Unlike most of Shakespeare's other history plays, there's very little actual fighting at the end of Act IV, when the Battle of Agincourt occurs. Further, Branagh's battle scene in the film brings to bear significant historical knowledge that Shakespeare could not use, plus several contemporary film technologies, including slow motion, to immerse his spectators in the fog, mud, and gore of imagined medieval warfare. Historians have known for a while that the longbow was the key to English victory at Agincourt and Branagh shows hundreds of arrows raining down from the skies on the French advance—arrows that decimated their historical cavalry and infantry. In short scenes of hand-to-hand combat, spectators sense the immense weight of the heavy broadswords, watch as arms, legs, and torsos brace and struggle against each other, catch a glimpse of Nym dying, stabbed in the back when he tries to steal a Frenchman's purse after killing him, and sees naked aggression on the King's face as he rushes against a foe—all of it in a rain-and blood-soaked field. The ten-minute battle seems an eternity of brute force and pain; I was exhausted from watching it.

Although the battle is an abrupt contrast to the soaring camaraderie of Harry's St. Crispian Day speech only minutes before, its main effect is to put English muscle, bone, and **altruism** behind the politics of brotherhood that the King preaches. Branagh's Harry begins informally, even humorously, smiling as he tells his cousin Westmorland that he likes the five-to-one odds in men that favor the French in the coming battle because, "The fewer men, the greater share of honor" (IV, iii, 22). Harry begins the speech on the ground, after dismounting, but he climbs to his left, ascending onto rocks, a cart, and eventually onto a promontory where he may view more of his army, still in dirty grey and brown costumes in contrast to the King's bright red and blue. The effect is to watch Harry literally rising to the occasion; the higher he climbs, underscored by thrilling music, the higher and more oratorical his speech becomes. As he continues talking about honor, it becomes clear during his ascent that honor is no longer a badge of individual courage, but a pledge to the fraternal right to die together: "We would not die in that man's company / That fears his fellowship to die with us" (IV, iii, 38–39). What began as a kind of mathematical joke (5 to 1 odds more than double the honor of 2 to 1 odds) ends in the recognition that soldiers have a right to fight together and die for each other if they commit to doing so. Will they make this commitment to **altruism** now is the implicit challenge of the moment.

Then Harry seems to change the subject when he paints a picture of the fun and honors his soldiers who return "safe home" will reap in old age when they celebrate today, the "Feast of Crispian," with their family and friends. At this, the narrative climax of the play, Harry imaginatively turns each soldier into an aged storyteller, who happily remembers and recounts, with embellishments, "What feats he did that day" (IV, iii, 51). The King even helps them to cast their stories with the names of those present in the crowd, "Familiar in his mouth as household words – / Harry the King, Bedford and Exeter," and other noblemen of name (IV, iii, 52–54). And we see many of those faces among the listeners in the film. Finally:

> This story shall the good man teach his son;
> And Crispin Crispian shall ne'er go by,
> From this day to the ending of the world,
> But we in it shall be rememberèd –
> We few, we happy few, we band of brothers,
> For he to-day that sheds his blood with me
> Shall be my brother. Be he ne'er so vile,
> This day shall gentle his condition.... (IV, iii, 56–63)

With these words, the King offers to join in a kind of communion, in this world and the next, with his soldiers. Both parts of the speech, the exhortation to cooperate together in committing to fight to the death with the King and the promise that all soldiers who do so will be "rememberèd" until "the ending of the world," fit together. Consequently, whether we win to tell our war stories to our sons or die together trying, we shall be remembered forever.

The key to eternal remembrance and glory, God's as well as man's, is the battle's link to the Feast of Crispian. According to Christian myth, the Feast celebrated the martyrdom of two Roman brothers, Crispin and Crispinian, both shoemakers, who journeyed to pagan France before Rome was a Christian empire to propagate the faith and were killed for preaching the Gospel. Harry's "we band of brothers" invokes the claim and urgency of **altruism** by extending the fraternal relationship of the two cooperating brothers to all English soldiers fighting with the

King and promises each eternal renown, "be he ne'er so vile"—be he even a simple shoemaker. By aligning himself with the martyrdom of the brothers, each warrior shall "gentle his condition" in the memories of God and man.

This promise, this symbolic link to martyrdom and memory, provides a much more gratifying answer to soldiers like Williams among the English troops than the theological niceties the King put forward to him before the dawn. Williams was specifically worried about the fate of all the chopped off "legs and arms and heads," prevented from a proper Christian death in battle, that might wander until the end of days between Heaven and Hell. The fear of unredeemed body parts haunts *Henry V* until the St. Crispin's Day speech, when it is resolved. For the most part, Shakespeare handles the problem comically, as when Mistress Quickly comments on Falstaff's death. After assuring the dying knight that there was no need yet to think of God, she put her hand under the bedclothes to feel first his feet, then his knees, "and so upward and upward, and all was as cold as any stone" (II, iii, 24–25). Katharine, the French princess destined to wed Harry after Agincourt to seal the peace between England and France, has a scene with her lady-in-waiting in Act III about the English names for body parts that also plays teasingly with sexual inuendo. After learning and mispronouncing English words that she mixes up with French sexual terms, Emma Thompson as the Princess and her maid collapse together in embarrassed, maidenly laughter. The point, of course, is that the same body parts that can generate love and life will one day die and, whether connected or not to the rest of a corpse, await the judgment of God.

After recounting the number of dead on the battlefield—over ten thousand French and around thirty English—the King reaches the conclusion that God was on the side of England. Harry's decision to link the outcome of the battle to the Feast of Crispian hovers in the background here and appears to justify this obvious conclusion. A few lines later, the King says simply, "God fought for us" and he asks that "Non nobis" and "Te Deum" be sung in thanks and celebration. Composer Patrick Doyle's plaintive rendition of "Non nobis" ends the scene. It is sung at first by a few soldiers, taken up by an off-stage choir, and climbs to full orchestration and majesty when the King, who has been carrying the dead body of one of the English boys in a long tracking shot across the battlefield to a pile of corpses to be buried or burned, puts the lad down gently and kisses him. Ceremony reigns in the final act of the play when Harry is courting

Katherine during the peace negotiations between England and France. Harry talks of holding "hands" and kissing "lips," but the conversation stops well short of impropriety and never approaches dead body parts.

2.4 Politics, Social Cohesion, and Altruism

What might this explanation based on enactive approaches to **empathy**, human **emotions**, and **narrative** teach us about the relationship between the politics of *Henry V* and the social cohesion of its audience, both in 1599 and 1989? Recall that, according to Whitehouse's coding practices for discovering such dynamical patterns, both performances are already close indeed to the religious rituals of the early agricultural period that preceded them in terms of ritual specialists, audiences, and checks on orthopraxy and orthodoxy. Still to be analyzed are the **euphoric** and **dysphoric** elements of the play and the film, plus the content of both performances insofar as they relate to matters of social cohesion and social–political complexity at the different times of their performances. I will first examine the euphoric and dysphoric elements likely present for both audiences in 1599 and 1989, then try to assess how both performances touched on matters of social cohesion and concern in their very different historical eras.

Unfortunately, Whitehouse invites his coders to perpetuate an outdated distinction between "active" ritual specialists and "passive" audiences with regard to the euphoric-dysphoric spectrum. An enactive approach to what happens in rituals and dramas, however, alters this sharp dichotomy. While it has always been true that performers, both shamans and actors, must initiate interactions with those assembled as spectators, the audience for rituals and dramas is by no means passive. In addition to moving their eyes, attuning their ears, and adjusting their bodies to get a better view of the performer's faces, spectators always neurologically mirror the intentional movements of the performers when the ritual or drama starts the processes of simulation and empathy and registers emotions for the making of meanings. Our species coevolved for social interaction, not passive reception.

Whitehouse's general question about the kinds of actions initiated by the performers and whether these simulations will likely initiate euphoric or dysphoric interactions, however, remains relevant. His list for coding cognitive and performative dysphoric elements includes "fear," "disgust," and "risk of death," along with "fasting," vigil/sleep deprivation," and "mutilation." In contrast, euphoric elements feature a range of positive emotions along with "feasting," "sex," "dancing," and "singing." Clearly, the primary actions and emotions of *Henry V*, from the decision to go to war through the Battle of Agincourt, fall on the **dysphoric** side of the spectrum. To be sure, there is what can rightly be called some comic relief from aggression, fear, disgust, and the threat of mutilation, but the play as a whole is primarily meant to rouse and then resolve dysphoric anxieties. Even the comic allusions to body parts collude with the threat of mortality. Although the audience in 1599 would have caught most of the allusions to these dysphoric elements through the language of Shakespeare's play, Branagh's film in 1989–90 also found disturbing visual imagery to awaken these same fears some four hundred years later.

In Elizabethan London, Shakespeare invited his audience at the Globe theater to process and resolve these dysphoric fears primarily through a commitment to **altruism** and Christian faith. Knowing that the desire for a Christian death followed by eternal salvation and a reputation for honor and bravery among one's tribe animated all of the souls in his audience, Shakespeare wrote the St. Crispin's Day speech as the climax and turning point of the play. The placement and likely success of the speech in 1599, of course, had significant implications for the probable boost that performances of the play gave to the social cohesion of London society toward the end of Queen Elizabeth's reign. When fellow soldiers believe themselves to be a "band of brothers," **altruism** heightens their own empathy and cooperation and the audience's social cohesion in the face of danger. As Whitehouse and Turchin have demonstrated, altruism reliably flows from dysphoric, **identity fusion rituals,** rendering *Henry V* a distant descendant of this ritual type. The hormones of altruism (primarily oxytocin) transform the numbing fear of death into **cooperation** and heighten tribalistic aggression.

Because **altruism** is a universal "attractor" of positive emotions often leading to increased social cohesion, successful modern productions of *Henry V* on stage as well as film can also be expected to generate what might be called a "we are all in this together" response. These feelings continue to be likely even though many spectators today no longer participate in identity fusion rituals and have no faith in or institutional affiliation with Christianity. For most modern spectators, the play's many links to Christian mythology are unlikely to move them to agree that the English won because God was on their side. The emotional appeal of altruism, not any particular ideology, probably accounted for the response of most spectators in 1989. Because ideological analysis of one sort or another remains a primary mode of critique among humanists, the shift from those methodologies to evolutionary explanation is a significant one. As we will see, ideology is still important in historical cultures, but it is more likely a secondary effect rather than a primal cause.

For other primal causes, we can turn to the political predilections of our species that have already been discussed. These, too, have helped to shape responses to significant public acts of altruism, often through language that reflects ideological preferences but is based on more fundamental realities. In the first chapter, I examined the coevolved tensions in matters of governance and authority between **equality** and **hierarchy**, plus **prestige** and **domination,** and also the ongoing conflict between varieties of **tribalism** (classism, racism, and ethnocentrism, etc.) and **xenophilia** (the coevolutionary desire to expand in-groups). A thorough examination of *Henry V* in the context of its 1599 and 1989 English cultures, the second clearly more complex than the first, would require this investigation.

Rather than launching a thorough-going analysis, however, I will limit my comments to three insights related to equality, xenophilia, and tribalism as examples of this possible discussion. I found it interesting that *Henry V* moves about as far as Shakespeare could probably go within the constraints of early modern social and economic hierarchy to emphasize the equality of all men, despite social rank, before God. The affirmation that, on the brink of mortality, all individuals, be they "ne'er so vile," should be considered as gentlemen in the eyes of God was certainly an appropriate Christian message (and clearly the right thing to say to medieval soldiers before battle), but the message also legitimated a notion of equality in the eyes of God with the potential power to frighten the mighty. Second, Branagh's embrace of xenophilia in his sympathetic treatment of Fluellen, plus the Scots and the Irish in Harry's army, was likely meant as a riposte to Thatcher's interest in emphasizing asocial, neoliberal policies, her backing a tough stand against the IRA, and her desire to break the power of Labor's solidarity with immigrant groups and with Scots who desired some form of independence from the UK.

Third, reinforcing the first of these considerations but not the second, the emotive power of **altruism** would tend to affirm the unity of all mortal Christians before the throne of God, but also deny the power of xenophilia to trump nationalistic tribalism. In the context of 1989, most Brits probably saw the film (and Branagh's stage production before it) as vindicating a resurgence of Rule Britannia in the Falklands War that occurred earlier in the decade. An interest in celebrating English imperialism also played a likely role in the original performances of the play in 1599. In fact, at the start of Act V, the Chorus refers directly to the likelihood of English military triumph, under the leadership of Robert Devereux, second Earl of Essex, against the Irish rebels he was sent by Queen Elizabeth to vanquish in that year. (Devereux failed and returned to London in disgrace.) All of these are local, historically specific situations partly shaped by cultural ideologies. But all of them also worked within universal patterns—pressures for equality, xenophilia, and tribalism—that emerged through coevolution. How the universal constraints of our coevolution shaped the major contours of world history and significant performances is the subject of Part II of this book.

References

Atran, Scott. 2002. *In Gods We Trust: The Evolutionary Landscape of Religion*. New York: Oxford University Press.
Boyd, Brian. 2009. *On the Origin of Stories: Evolution, Cognition, and Fiction*. Cambridge, MA: Belknap Press.
Boyer, Pascal. 2001. *Religion Explained: The Evolutionary Origins of Religious Thought*. New York: Basic Books.
Boyd, Robert, and Peter J. Richerson. 1985. *Culture and the Evolutionary Process*. Chicago: University of Chicago Press.
Colombetti, Giovanna. 2014. *The Feeling Body: Affective Science Meets the Enactive Mind*. Cambridge, MA: MIT Press.
Damasio, Antonio. 2003. *Looking for Spinoza: Joy, Sorrow, and the Feeling Brain*. New York: Harcourt.
Grodal, Torbin. 2009. *Embodied Visions: Evolution, Emotion, Culture, and Film*. New York: Oxford.
Hogan, Patrick C. 2018. *Literature and Emotion*. New York: Routledge.
Hutchins, Edwin. 1995. *Cognition in the Wild*. Cambridge MA: MIT Press.
Johnson, Mark. 1987. *The Body in the Mind: The Bodily Basis of Meaning, Imagination, and Reason*. Chicago: University of Chicago Press.
Lanman, Jonathan A. 2007. How "Natives" Don't Think: The Apotheosis of Overinterpretation. In *Ritual, Anthropology, and Cognitive Science*. ed. Harvey Whitehouse and James Laidlaw, 105–32. Durham: North Carolina Academic Press.
McConachie, Bruce. 2015. *Evolution, Cognition, and Performance*. Cambridge: Cambridge University Press.

Seshat/Methods/Code book/Ritual Variables. (n.d.). Retrieved 8/6/2019. From https://www.seshatdatabank.info/methods, (n.p.).

Shakespeare, William. 2017. *Henry V* (The Pelican Shakespeare). Kingston, Jamaica: Pelican.

Smith, Murray. 2017. *Film, Art, and the Third Culture: A Naturalized Aesthetics of Film*. Oxford: Oxford University Press.

Thompson, Evan. 2007. *Mind in Life: Biology, Phenomenology, and the Sciences of Mind*. Cambridge: MA: Belknap Press.

Throop, C. Jason and Zahavi, Dan. 2020. Dark and Bright Empathy: Phenomenological and Anthropological Reflections. *Current Anthropology* 61 (3) (June 2020): 283–92.

Turner, Mark. 1996. *The Literary Mind: The Origins of Thought and Language*. New York: Oxford.

Whitehouse, Harvey. 2004. *Modes of Religiosity: A Cognitive Theory of Religious Transmission*. Lanham, MD: Altamira Press.

Whitehouse, Harvey, P. François and Turchin, P. 2015. Social Evolution Forum: The Role of Ritual in the Evolution of Social Complexity; Five Predictions and a Drum Roll. *Cliodynamics: The Journal of Quantitative History and Cultural Evolution*, 6 (2): 199–216.

Whitehouse, Harvey, and J. Lanman. 2014. The Ties That Bind Us: Ritual, Fusion, and Identification. *Current Anthropology* 55 (6): 674–695.

PART II

Political Universals in History and Performance

The three chapters of Part II examine key ritual and theatrical performances and their politics from the deep history of the Pleistocene until around 1960. I am interested in significant rituals and dramas from major and increasingly complex cultures in world history and the universal political orientations they share—performances in significant polities from hunter-gatherer, agricultural, and capitalistic eras. The point is not to "prove" that altruism, tribalism, equality, and the other indicators of our political nature have guided the deep and shallow history of our species. No historical overview, however detailed, could validate the contours of human nature, in politics or any other area of behavior. That's a matter for genetic theory and rigorous experiments to demonstrate. What concerns me in Part II is the coevolution of these indicators (in Chapter 3) and their changing importance in world history (in Chapters 4 and 5), as warfare and increasing social complexity during the agricultural and capitalistic eras kicked up different possibilities for human performances, which, in turn, helped to shape the dynamics of larger historical processes.

Overall, the major performances I discuss in Part II move from a Late Pleistocene-appropriate ritual in Africa to *The Life of Galileo*, by Bertolt Brecht, in 1947. Along the way, I also deploy Whitehouse's analytical framework to an ancient Greek performance that is part ritual and part theater, a bunraku action-adventure puppet play in early eighteenth-century Japan, and to the Mozart-Da Ponte opera, *The Marriage of Figaro*, in enlightenment Vienna. Although a few other euphoric and dysphoric rituals and dramas will claim my attention during these chapters, these five performances deserve some advance billing. As will be evident, the political nature of our species—the tensions between domination and prestige, tribalism and xenophilia, equality and hierarchy, and cooperation and competition—shaped all five.

CHAPTER 3

Coevolution in the Pleistocene

Our genus, *Homo*, lived over 95% of its time in the Pleistocene Epoch. During that geological period, according to human evolutionary biologist Joseph Henrich, our species "evolved an addiction to culture" (Henrich 2016: 3). This is not to say that we share little with other primates. Indeed, as I will show, our emotional intelligence (and consequently the basis of our proclivity for culture) builds upon commonalities we share with chimpanzees and bonobos, our near relations on the evolutionary tree. Coevolution fitted our bodies and brains to learn a particular range of cultural practices in a changing environment with other enculturated hominins like ourselves. As noted in the last two chapters, these cultural predispositions inherited and learned during the Pleistocene included cooperation, empathy, and an aptitude for following social norms. They ranged between domination and prestige as the basis for the exercise of legitimate leadership, stretched between hierarchy and egalitarianism, involved tribalism and xenophilia with regard to group dynamics, and were partly transmitted through rituals that spanned the dysphoric-euphoric spectrum.

3.1 Coevolution in the Early and Middle Pleistocene

Lasting from 2.5 million until 13,000 years ago (ya), the Pleistocene Epoch alternated between periods of glaciation, which froze much of the Earth's water, and periods of warmth and thaw, when more water filled the seas, lakes, and rivers, which improved the evolutionary chances for most land-based flora and fauna. Some of the shifts between these periods were relatively rapid, leading to large storms, droughts, and fires that caused substantial plant and animal die-offs. Our ancestors stayed in Africa for most of the epoch, probably avoiding the worst of the weather, but began to venture into the vast Eurasian landmass about 800,000 years after the Epoch began.

The earliest fossil remains of *Homo habilis* (appropriately translated as "handyman") have been dated at 2.6 mya, a mere 100,000 years before the beginning of the Pleistocene. This first member of the Homo genus lived in the plains rather than the trees of Africa, practiced bipedalism, had a smaller digestive tract than her predecessors, and also used her hands to shape better stone tools than had her precursors on the family tree. Animal bones found at early *H. habilis* sites indicate that these hominins were scavenging and butchering the remains of giraffes, buffalos, and even occasional elephants after larger predators had killed them. The fish bones and turtle shells at some of the later sites also suggest that this hominin had access to more sources of protein than his predecessors. By around 2 mya, the remains from one site in Africa support the scientific hunch that more protein in the diet led to hominins with bigger brains, especially in the frontal lobes. These areas of the brain would later support speech, gesture, and more executive control over tool and symbol use.

The evolution of our genus began to accelerate with the emergence of *H. erectus*, around 1.8 mya. Benefitting from the sharper tools of *habilis*, *H. erectus* continued to downsize the jaws, teeth, stomachs, and intestines of his predecessors—apparently no longer necessary for processing raw food—for more brainpower. This was probably due to the use of fire for cooking. Just as important, there is substantial evidence for cooperation among the early members of these hunter-gatherer bands. Anatomical changes to the new species not only permitted a fully erect posture and faster running, but also altered the muscles of the shoulders and wrists to allow for much more accuracy and speed in throwing stones and spears. Both would have facilitated the slaying of small game by hunting parties and more efficiency in group scavenging. The new tools invented by these hunter gatherers also suggest that these bands were sharing food with each other. *H. erectus* was also a species on the move. Soon after its emergence in Africa, bands of *H. erectus* migrated into Asia, moving as far as the Caucuses in present-day Russia. This mobility strongly suggests some reliance on learning culture to enable *erectus* bands to cooperate in adapting to novel environments.

Evolutionary anthropologist Sarah Hrdy makes a strong case for another kind of public cooperation among these early *erectus* bands—cooperative child care, also known as alloparenting. Hrdy uses neuroscience, primatology, archaeology, and social and developmental psychology in her *Mothers and Others* (2009) to conclude that "there emerged in Africa a line of apes that began to be interested in the mental and subjective lives – the thoughts and feelings – of others, interested in understanding them. These apes were markedly different from the common ancestors they shared with chimpanzees, and in this respect they were already emotionally human" (Hrdy 2009: 31). Part of the reason *erectus* hominins substantially differed from previous primates was that their babies required more attention and interaction for a longer time than chimp infants. Pressed by early cultural evolution to increase the amount of time that babies could take to learn their culture, genetic evolution had increased the size of an infant's brain at birth and the length of childhood, with the result that they were born nearly helpless with regard to their physical survival skills, but ready to learn immediately from their caregivers through empathy and direct instruction. Coevolution would continue to expand our species' brain size and childhood for the next million and a half years.

According to Hrdy's evidence, alloparenting in the early Pleistocene involved group cooperation among *H. erectus* adults and older children in the same band—primarily women and older girls—for the care and provisioning of infants and the young. As Hrdy explains, "Both before birth and especially afterward, the mother needed help from others; and, even more importantly, her infant would need to monitor and assess the intentions of both his mother and these others and to attract their attentions and elicit their assistance in ways no ape had ever needed to do before" (Hrdy 2009: 31). Indeed, other great ape mothers hold fast to their infants, entrusting them to no one out of fear for their lives. By involving a band's alloparents in imitating and improving upon each other's "baby sitting" methods, alloparenting likely led to Pleistocene versions of best practices in child-raising in each band. This imitative cooperation based on altruism probably improved the chances of survival for *H. erectus* infants and children.

Effective alloparenting depends upon what evolutionary anthropologist Michael Tomasello calls "shared intentionality." This ability entails: "(i) The cognitive skills for creating joint intentions and attention (and other forms of common conceptual ground) with others; and (ii) the social motivations for helping and sharing with others (and forming mutual expectations about these cooperative motives)" (Tomasello 2008: 73). Whereas chimps, baboons, and other great apes appear to be incapable of acting with shared intentionality, human babies today generally learn these skills at around nine months of age. When our infant ancestors began to practice shared intentionality cannot be known with any certainty, but *H. erectus* babies were probably born with some genetic proclivity for this ability that natural selection and cultural practice likely strengthened during the mid-Pleistocene. Like alloparenting generally, shared intentionality would have improved cooperation among the hunter gatherers practicing it and exerted coevolutionary pressure to increase the hormones for altruism in the genetics of the band.

Although there is some evidence that adult *H. erectus* males occasionally participated in alloparenting, most no doubt learned collaboration and friendship through hunting together. Again, this was a distinct improvement over the typical independence of other Pleistocene apes during their hunting forays. As with alloparenting, shared intentionality and mutual empathy were important components of male hunting, which consisted primarily of scavenging food from larger predators and killing small mammals. Effective scavenging required collaboration among several men. A group armed with spears and rocks was needed to keep the killer of the animal occupied while a few of the hunting party worked quickly to strip as much meat from the prey as they could carry. Over many generations, as with alloparenting, the empathy involved and the prosocial emotions of sympathy, pride, gratitude, loyalty, and occasional episodes of shame probably forged each *H. erectus* group of hunters into a Pleistocene "band of brothers."

According to biological anthropologist Terence Deacon, by the middle years of the Pleistocene, the *Homo* genus experienced an evolutionary bottleneck in the growth of their cooperative sociality. Without the ability to legitimate the social rights and obligations of long-term sex partners, he believes, the unstable cohesion of these *H. erectus* bands could easily fragment through male rivalries and sexual jealousy. States Deacon: "The need to mark these reciprocally altruistic relationships arose as an evolutionary adaptation to the extreme instability of the combination of group hunting/scavenging and male provisioning of mates and offspring" (Deacon 1997: 401). What was required, according to Deacon, was the social validation of sex contracts—the performance, that is, of a kind of Pleistocene proto-marriage. This had not been a problem for early chimpanzees, whose genetics and learning fitted them for contests of dominance in every aspect of their lives, including sexuality. It is clear as well that, due to the predominance of alloparenting and cooperative hunting among our *H. erectus* ancestors, there would have been pressure on the males to share the spoils of the hunt among the entire band, not just with sexual partners and offspring. Nonetheless, bands probably dealt with this problem in a variety of ways and the difficulty of achieving consensus on the matter of heterosexual coupling could—as we know from present-day experience—kick up many other problems for the emerging cooperative sociality of the band.

Were there social rituals to legitimate the proto-marriages of *H. erectus* males and females during the middle years of the Pleistocene, as Deacon assumes? This is probably impossible to determine. Over time, however, interactions between male and female groups in the same band probably worked to make relations between the sexes more egalitarian. Although the types of labor dividing the sexes likely changed very little, Pleistocene women apparently gained increasingly equal power within their bands. Among the factors causing this increase were likely the growing importance and control of fire for cooking (probably attained long before 1.5 mya, the likely date for the first evidence of fire use), environmental factors related to a new ice age in the Pleistocene that limited the success of male hunting, and innovations in hominin weaponry. Regarding their weapons, for the first time, groups of women could potentially gang up on an abusive alpha male in the band and kill him. Whether such gender revolts occurred often cannot be known, but most adults in the band probably understood that they were possible; by itself, this knowledge likely helped to level the field of power relations between the sexes. The first social victory of female adults, whether achieved violently or peacefully, probably had to do with restraining male physical threats in order to gain greater control over their sexuality. This "reverse dominance," as anthropologist Camilla Power and others have termed it (Power 2014: 196), allowed a female collective to deter and dominate individual males that sought to dominate them. What likely began in the private sphere of hunter-gatherer life emerged into the public sphere of the band and shifted power relations among the genders. *H. erectus* rituals of proto-marriage, of course, may have provided another public restraint on male sexual behavior.

With changes in gendered power relations during the middle Pleistocene came the possibility of prestige as a mode of attaining public status and political authority within hunter-gatherer bands. As Henrich, who did extensive original research on the evolution and psychology of prestige, explains in *The Secret of Our Success*, prestige became important "once humans became good cultural learners [and] needed to locate and learn from the best models.... As a consequence, humans reliably develop emotions and motivations to seek out particularly skilled, successful, and knowledgeable models and then are willing to pay deference to those models in order to gain their cooperation (pedagogy), or at least acquiescence, in cultural transmission" (Henrich 2016: 118–119).

The attainment and practice of authority through prestige differs crucially from systems of authority based on domination, which our species inherited from our primate relatives. Human physical interaction varies significantly in these two systems of public authority. Male domination during hominin times and continuing today is based primarily on threat and coercion. In the presence of a dominant alpha male, inferiors of both sexes generally avert their eyes and keep their distance to avoid random aggression. In contrast, females and males gain prestige from others through persuasion and deferential agreement, not threats. Responding to the aggression, egotism, and self-aggrandizement of dominators, inferiors typically respond with fear and shame. In sharp contrast, the prosocial actions and generous teaching of the prestigious may inspire admiration, awe, and respect—but rarely fear—from their followers. The ingredients for authoritarian leadership would have been present when nearby bands of *H. erectus* "others" threatened traditional hunting and gathering grounds during the middle and late centuries of the Pleistocene. But the most successful hominin cultures apparently chose leadership modes of prestige rather than domination. On balance, leaders with the authority of prestige were probably more effective in generating group cooperation than those who practiced dominance. As Henrich explains, "The reason lies in our cultural nature." Because they are role models for others, prestigious leaders that behave with generosity and altruism "can increase the overall prosociality" of their groups (2016: 129).

Prosocial practices and emotions in the public sphere of our Pleistocene ancestors are a part of the answer as to when our species "crossed the Rubicon into a regime of cumulative cultural evolution, which has driven human genetic evolution ever since," says Henrich (2016: 280). In general, affirms E.O. Wilson, "Human behavior is determined neither by genes nor culture but instead by a complex interaction of these two prescribing forces, with biology guiding and environment specifying" (Wilson 2005: vii). Although this generalization remains true, the gradual coevolution of *H. sapiens* began with genes as the primary generators of our evolution, but gradually ceded causal primacy to culture. Henrich describes the past and ongoing interaction between genetic and cultural evolution as "*autocatalytic*, meaning that it produces the fuel that propels it.... As genetic evolution improved our brains and abilities for learning from others, cultural evolution spontaneously generated more and better cultural adaptations, which kept the pressure on for brains that were better at acquiring and storing this cultural information" (2016: 57.)

So, when did culture take the place of genetics in the evolutionary driver's seat? Henrich arrives at 1.8 mya as the rough date of the transition that put culture ahead of genetic evolution in causal importance. He dismisses those anthropologists who believe that one single event or innovation—such as the emergence of hands that could exercise a precision grip or the invention of cooking—might have suddenly allowed our ancestors to move from the slow lane of genetics into the fast lane of culture. Henrich argues persuasively that "with low-quality imitative skills, error-prone copying, social intolerance, and little or no teaching, accumulating and sustaining cultural adaptations would have been very delicate in the early days [between 6.5 and 2.5 mya]" of our evolution (2016: 280–281). Looking broadly at the evolution of stone tools, changes in hominin anatomy, and genetic evidence from ancient DNA—"stones, bones, and genes" (281)—Henrich finds some evidence of cumulative cultural evolution as far back as 3.4 mya, but underlines the emergence of *H. erectus* at 1.8 mya as the crucial turning point. In effect, the new species on the block was the result of both genetic and cultural adaptations for larger brains, faster speed, better tools, more accurate throwing ability, a wider range of vocal production, and a host of smaller changes that allowed culture to surpass genetics as the primary innovator of change. As we have seen, the new species came ready-made for prosocial public practices, altruistic motives and emotions in childcare and hunting, and modes of authority and leadership in the public sphere that would soon lead to further transformations.

3.2 Cooperative Cultures in the Late Pleistocene

Evidence from several sites indicates that *H. erectus* bands survived well into the next millennium, testament to their learning and adaptability to withstand several ice ages. Because their cultural innovations also drove faster genetic changes, these alterations gradually produced a new species, *H. heidelbergensis*, which emerged eight to seven hundred thousand years ago. As the name suggests, this descendent of *H. erectus* was first discovered near Heidelberg in Germany; the species apparently survived until twenty-five thousand years ago. *H. Heidelbergensis* bands probably passed into extinction only because two other species evolved directly from them to challenge their dominance—*H. neanderthalis* and *H. sapiens*. Spread across Africa and Eurasia, *H. heidelbergensis* were the first hominins to build shelters, mostly of rock and wood, and to hunt big game, with long, heavy spears. These women and men had larger brains than their predecessors and anatomical studies of their skulls and necks suggest that they were much better equipped for making and listening to articulated and musical sounds than any species before them. Because the two species that inherited most of their genes performed a variety of rituals to ease social conflicts, it is likely that *H. heidelbergensis* practiced some rhythmically infused ceremonies and work routines as a regular part of their culture.

H. heidelbergensis bands tended to be more populous than those of *H. erectus* hominins, primarily because their advances in weaponry, hunting techniques, and uses of fire (for protection against predators as well as cooking) required and could support more members. To survive ice-age winters, the bands living in Europe mobilized their members for large-scale hunts that targeted mammoths, Irish elk, and European lions, among other big game. Attacking from several sides of their prey, they used their long, thick spears to pinion the thrashing animal in place and jab it to death. Once the beast had died, members of the band—perhaps women and older children as well as men—had to quickly strip the carcass of meat and transport it back to their camp before other predators moved in to attack them and scavenge the remains. Perhaps they used primitive rhythm instruments—maybe sticks, bones, and hollow logs—to coordinate and speed the work. The many injuries found by anthropologists in

H. heidelbergensis skeletons attest to the life-threatening dangers of their mode of hunting.

Such demands probably helped to produce rigorous social norms among *hiedelbergensis* cultures that could be passed down over several generations. Social norms go beyond temporary agreements and cooperation among trusted friends. They hinge on the understanding that moral behavior is a matter of life-long collective commitment, publicly applicable to everyone in the band. Comparing human norms to chimpanzee rules for behavior, psychological anthropologist Michael Tomasello has noted that "whereas human norms are often variable across groups, gestures negotiating the relationship between dominant and subordinate individuals in chimpanzees are highly similar across different, unrelated populations," which suggests that genetics plays a larger role than culture in chimp behavior (Tomasello quoted by Haun and Over 2013: 83). In addition, humans usually care about punishing those who break social norms, even if the transgression does not affect them directly; chimps, however, rarely act in this way. Tomasello adds, "[T]he domains most universally covered by social norms across societies are those involving the most pressing threats to the group's cohesiveness and well-being; that is, those that bring out most strongly individuals' selfish motives and tendencies to fight: food and sex" (Tomasello 2016: 98). Evolutionary philosopher Kim Sterelny provides persuasive evidence that the ability to craft, follow, and enforce social norms is both a biological predilection and a culturally learned ability; all cultures today deploy social norms, although these norms vary widely throughout the cultures of the world.

Pressure for enforcing social norms to ensure public cooperation may also have led *H. heidelbergensis* bands to create stronger bonds of tribalism than had been possible before. Then as now, the ethnocentrism that comes with tribalism is learned in childhood and reinforced through rituals and other practices of sociality during early adulthood. When children begin to play with others, they join groups and learn appropriate beliefs and behaviors from their peers. These social groups, influenced as well by adult social norms, establish their own social pressures and punish outliers. As children grow up and are integrated into larger public groups, the need to recognize and separate "us" from "them" becomes increasingly important. Young adults also learn to rein in and discipline free riders, those who refuse to pull their weight in the collective effort to survive.

In addition to learning, it is likely that *H. heidelbergensis* was genetically predisposed to distinguish their in-group social identity from those "others" nearby who practiced different cultural habits. According to social identity theory, begun by Henri Tajfel in the 1960s, our species evolved to practice modes of social categorizing that worked to heighten differences between "insiders" and "outsiders." As well as giving people a sense of self-identity and esteem, one's social identity typically shapes relations with other groups and may result in prejudice and stereotyping. It is likely that genetics and learning combined in *H. heidelbergensis* produce social identities with strong social norms. This was so "for at least three immediately prudential reasons," concludes Tomasello: "to make sure others could identify them as in-group members, to coordinate with the group, and to avoid punishment, including threats to reputation" (Tomasello 2016: 100).

Given the importance of meat for survival, *heidelbergensis* bands likely enforced strict public norms of meat sharing. Because large animal kills could be rare and disabled hunters might go weeks without success, most bands probably tried to ensure that all of their members ate meat when they could. Contemporary hunter-gatherer bands typically rely on one of several normative codes to regulate the sharing of meat. In some bands today, for example, the "ownership" of the meat is transferred from the hunter to a third party, who is then trusted with its distribution. Food taboos, specifying, for example, that the hunter and his wife and children are allowed to eat only certain parts of the prey, is another way of facilitating wide distribution. Because meat is typically consumed publicly, around a communal campfire, band members can monitor and judge others for violations of meat sharing norms. These norm violators typically suffer a damaged reputation, which in turn can lead to gossip and social isolation. Paleoanthropologists believe that these and other contemporary techniques for the cooperative sharing of meat have been used by hunter gathers for hundreds of thousands of years.

The rough egalitarian practices for public meat sharing among *H. heidelbergensis* hominins probably helped to shape their norms of leadership and governance. Meat-sharing norms, for example, likely made it impossible for alpha males, even those who were excellent hunters, to keep all the meat they brought back from the hunt for themselves and their families. In their essay 'Zoon Politicon' in *Cultural Evolution*, Herbert Gintis and Carel van Schaik underline the continuing preference for modes of authority based on prestige rather than domination among contemporary hunter gatherers. A dislike of coercion is central to the politics of contemporary hunter-gatherer bands, state Gintis and van Schaik: "[They] share with other primates the striving for hierarchical power, but social dominance aspirations are successfully countered because individuals do not accept being controlled by an alpha male and are extremely sensitive to attempts of group members to accumulate power through coercion" (Gintis and van Schaik 2013: 36). Arrogant bullies and lying blowhards will be warned, punished, and occasionally ostracized from the band. If these strategies are unsuccessful, note Gintis and van Schaik, "the group will delegate one or more members, usually including one close relative of the offender, to kill him" (2013: 36). In the absence of traditional forms of hierarchy, these small bands generally govern themselves through persuasion and coalition building. The result, states one political scientist, is that small groups of individuals within these bands can often gain influence "through such prosocial currencies as the ability to mediate or organize defense, ritual, and exchange" (Wiessner quoted in Gintis and van Schaik 2013: 38). Governance by reverse–dominance coalitions probably remained the predominant form of politics among our genus from *heidelbergensis* times until the transition to agriculture.

In addition to the importance of social norms to our *heidelbergensis* ancestors for governing through coalitions of prestige, genetics also played a part. In his *The Social Ladder: How Inequality Affects the Way We Think, Live, and Die* (2017), social psychologist Keith Payne lays out the evolutionary case for egalitarian preferences among all primates, not just humans. Payne sites the work of primatologist Sarah Brosnan with capuchin monkeys, which demonstrates that these monkeys' "sense of fairness" is built into their DNA (Payne 2017: 21). "The discovery that capuchin monkeys [which last shared a common relative with *H. sapiens* about 25 million years ago] are averse to receiving unequal outcomes, much like humans, suggests that these tendencies are evolved rather than learned," states Payne (2017: 22). He adds that, if this is so, children at an early age—before social norms have directly shaped their behavior—ought to have the same predisposition toward fairness. And they do. In a series of experiments with three-year olds similar to those in which the monkeys participated, psychologists found that "the children became visibly upset when they received less than their partner [for completing the same task]. As every parent of preschoolers knows, they do not need to be taught that receiving the same amount is fair but receiving less is unfair" (22).

3.3 The Invention of Language

Despite the increasing complexity and norm-driven morality of *H. heidelbergensis* culture, there is no evidence that these hominins ever developed symbolic language. While it is probable that *erectus* and *heidelbergensis* hominins used gestures, pantomime, and simple vocal and whistle sounds to signal what could be communicated in such ways, linguists are clear that spoken language must be built upon combinatory symbols—that is, constructions that use a fixed number of phonemes to create multiple meanings—before it qualifies as a language. Further, language use involves what evolutionary linguist Daniel Dor calls "the systematic *instruction of the imagination*" (ital in original) (Dor 2014: 106). An epistemological addition to learning through experience, language, says Dor, "allows speakers to intentionally and systematically instruct their interlocutors in the process of imagining the intended experience, as opposed to directly experiencing it" (Dor 2014: 106). Group conversations concerning possible futures, for example, required language speakers to use a code that encouraged all of them to draw on their memories as well as their present knowledge in order to exchange novel ideas that only existed in their imaginations. While the extent to which our ancestors might have been able to use their proto-languaging skills to communicate about possible futures cannot be known, it is difficult to understand how such a simple question as "Should we hunt in a different forest tomorrow?" might have been communicated without language. Could *H. heidelbergensis* bands sustain such symbol-laden, imagination-dependent conversations? In Dor's understanding of language evolution, hominins practicing mimetic modes of proto-language "had to reach the limits of experiential communication and build communities complex enough, dependent enough on communication, and sophisticated enough in terms of collective innovation to begin the exploratory search for means of communication that could bridge the experiential gaps between the communities' members" (Dor 2014: 118). It took the emergence and flourishing of *H. sapiens* to perfect a communicative technology—symbolic language—that could make personal imagination available for public conversation.

Although the brains of the new species were only slightly larger on average than those of *H. heidelbergensis*, *H. sapiens* emerged "language-ready" around 300,000 ya, along with bodies that were equipped for producing and listening to sounds just as well as present-day humans. According to evolutionary cognitive scientist Chris Sinha, "Early language... involving lexically-based constructions and differentiated participant roles can be hypothesized to have emerged as the first original biocultural semiotic artifact of the language-ready brain 200–100 KYA [thousand years ago]... [E]volutionary modern languages (grammaticalized, morphosyntactically more complex, and with elaborated functional differentiation) probably date from 100 to 60 KYA" (Sinha 2014: 46). After that, as more complex social structures emerged, mythic narratives, religions, and other imaginative arts began to flourish.

H. sapiens inherited and improved upon all of the hominin predilections and practices involving prosocial emotions and social norms. Using the language of one's culture—conforming to the symbolic vocal combinations preferred by others in your neighborhood—is, of course, yet another instance of normative behavior. In an article entitled, "Forever united: the co-evolution of language and normativity," Ehud Lamm notes that several aspects of language are overtly normative: "(1) symbolization, which involves accepting arbitrary signs as appropriate labels; (2) most pragmatic phenomena, such as speech acts..., grasping communicative intent, and displaying conversational skills; and (3) understanding the normative context of discourse, which can transform questions into commands, requests into demands, and so on" (Lamm 2014: 268).

Turning to the coevolution of norms and language between 200 and 100 kya, Lamm notes that changing ecological conditions and competition among groups of *H. sapiens*, *H. neanderthalis*, and *H.heidelbergensis* put evolutionary pressure on the early humans for innovations in communication that could accommodate these challenges. In addition, the spread of *H. sapiens* cultures and the growing size of their groups required that language serves as a marker to differentiate the emerging tribes and their bands from each other. By 150 kya, most ethnolinguistic groups of our species probably ranged from around 800 to 1,200 people, the majority of whom were not directly related to each other. Over time, normative public processes improved upon, elaborated, and clarified the language of norms and morality within each band and tribe. It is also likely that language empowered those who sought authority through prestige, rather than domination. Now the prestigious could explain their ideas about the past and future, not just engage in extended pantomimes of noise and gesture about present concerns. They could also converse with others about what they believed to be truths about human experience and the world; the origins of philosophy and science begin with language.

The social-political effects of oral communication appear to follow the general outlines of communication practices predicted by Marshall T. Poe for Pleistocene hunter-gatherer cultures in his *A History of Communications*. Because oral communication was open and diffused among all adults in each language-based band, the cultural norm favored by such accessibility, says Poe, tends to be "egalitarianism" (Poe 2011: 58). At the same time, the general lack of privacy in a small community of speakers, especially within each male and female group, would tend to produce public rather than private modes of communication. An oral culture lacking any means to off-load cultural memory through writing needs traditional ritual actions to "remember" important events; consequently, wisdom will tend to reside in the conventional practices and sayings of tribal elders. In some tension with this norm, the dialogic attribute of an oral network will tend to produce social practices that are "democratized" and "amateurized," according to Poe (2011: 58). As we will see in the next chapter, the evolution of writing during the agricultural period, which was limited to and controlled by elites, helped to solidify the authoritarian tendencies of the archaic states of the period, fatally undercutting the loose solidarities of hunter-gatherer governance by mixed-gender coalitions.

The triumph of normative language and the larger populations of hunter-gatherer bands and tribes no doubt increased the importance of tribalism among the early *H. sapiens*. According to Avi Tuschman, there are three politically interrelated components to tribalism: ethnocentricity (already discussed), religiosity, and sexual (in)tolerance. "Individuals with this cluster of traits," says Tuschman, "tend to have political views on the right. On the other end of the spectrum, attraction to out-groups (xenophilia), secularism, and higher sexual tolerance are well correlated with one another and with political views on the left" (Tuschman 2013: 63). The coevolutionary consequence of each of these three groups of traits is that those who share them are more likely to choose to reproduce our species with another of the same orientation; both tribalistic people and cosmopolitan people tend to mate with their own kind.

Tuschman also reports research on contemporary ethnoreligious conservatives that demonstrates that these populations around the world are "less tolerant toward all forms of sexuality *other* than sexual reproduction within their group's sanctioned form of marriage" (2013: 106); the more tribalistic a group, the more it probably objects to LGBTQ expressions of sexuality. Extremely conservative cultures are also more likely than only moderately tribalistic ones to ensure that their young women marry the "right" man by controlling a young woman's freedom in "(1) sexually attracting male strangers in public, (2) choosing a spouse, and (3) divorcing" (124), notes Tuschman. The ultra-gendered culture of contemporary Saudi Arabia is a good example of the interlocking confluence of these practices. While there are significant historical reasons for why various populations adopt differing cultures of tribalism or cosmopolitanism regarding ethnocentrism, religiosity, and sexual tolerance, these patterns can change over a few generations, primarily due to new mating habits forced by war or changes in migration. As Tuschman reports,

> It's difficult to overemphasize the magnitude of the shift in gender equality that has taken place in many developed countries. In the United States, for example, one of the largest changes ever recorded in public opinion concerns approval 'of a married woman earning money in business or industry if she has a husband capable of supporting her.' In 1937, only 18 percent of Americans accepted this proposition; by 1990, 82 percent approved. (Tuschman 2013: 212)

How liberal or conservative sexual attitudes and behaviors might have been among different tribes toward the end of the Pleistocene cannot be known, but it is probable that the coevolutionary dynamics controlling the parameters of these social actions were much the same then as now. The new uses of language, however, probably increased the ability of our species to wage war. Just as our ancestors occasionally continued to elevate dominant males into positions of authority as a result of our genetic inheritance from other great apes, we seem to have inherited a predisposition for warfare from them as well. Contemporary troops of chimpanzees organize warring raids against neighboring bands of their species in order to expand their hunting and gathering territory. Chimp group behavior is "particularly interesting," says Henrich, "because they provide a ready model for what the common ancestor of humans and chimpanzees might have been like. If modern chimpanzees do it, it's plausible that the common primate ancestor we share with chimpanzees might have done it also" (Henrich 2016: 170). One study of chimpanzees in Uganda found that adult male chimps from a large troop were able to expand their territory by over twenty percent through occasional raids across the borders of a smaller neighboring troop during a ten-year period. These attacks, resulting in 21 deaths, forced the retreat of the smaller troop and permitted the females and children of the larger troop to occupy the new territory.

Were such raiding parties and murderous outcomes a common feature of our evolving species soon after we parted company with our common chimpanzee ancestor over six million years ago? That is more difficult to know. Most of the evidence for massacres among groups of our ancestors is more recent, occurring after the emergence and success of *sapiens*. It may be that groups of *H. erectus* and *heidelbergensis* were generally spread more thinly in Africa and elsewhere than chimp troops are today, leaving more space between the hunting and gathering territories of potentially rival bands. On the other hand, the first clear evidence of cannibalism—human teeth marks on human bones—dates from around 800,000 years ago, soon after *heidelbergensis* first appears. Perhaps the tightening of social norms in *heidelbergensis* culture that led to more cooperation within the band also sharpened us–them distinctions with outsiders to a point that encouraged the public eating of enemy others.

Warfare, however, was not the only way of dealing with intergroup competition. Migration would have been an alternative for some individuals and bands. Following in the steps of their ancestors, *H. sapiens* began moving out of Africa about 180,000 ya; they spread throughout Eurasia, peopled the islands of South Asia and Australia, and eventually advanced across a land bridge into the Americas. (Until recently, most anthropologists dated *sapiens* emigration out of Africa at around 60,000 ya, but discoveries in Israel have substantially pushed back that date). During these decades of (very) Late Pleistocene migration, it is probable that many hunter-gatherer bands decided to move to new territory rather than fight their neighbors for older lands that were already being depleted of easily hunted game. Henrich remarks that "differential migration," which occurs when individuals in a failing band decide to migrate to a more successful one, is another alternative to warfare in intergroup competition (168). *Sapiens* bands that were more successful than others in finding and securing new hunting grounds would have been a magnet for such individual migration and these more fit bands would have had several inducements to incorporate new members.

Henrich also lists "prestige-based group transmission" as another peaceful means of besting others in intergroup competition while also avoiding war. He builds on Boyd and Richerson's coevolutionary research as well as his own to offer this explanation: "Because of our cultural learning abilities, individuals will be inclined to preferentially attend to and learn from individuals in more successful groups.... This causes social norms, including ideas, beliefs, practices (e.g., rituals) and motivations, to flow via cultural transmission from more successful groups to less successful groups" (Henrich 2016: 168). Henrich offers as an example the elders of the New Guinea village of Irakia, who realized that the best way to improve their lagging pig production was to adopt significant parts of the culture of a nearby tribe, whose pig production was substantially better than their own. Accordingly, after watching the cultural practices of their neighbors, the Irakia elders promulgated several measures related to pig husbandry, which included feeding them more food, allowing pigs to roam more freely among village gardens, and even rituals that got them singing, dancing, and playing their flutes for their pigs. The anthropologist who observed this transfer of knowledge reported a short-term increase in Irakia pig production. This is not surprising, says Henrich. He notes that "the copying of institutions and rituals from more successful groups is commonplace" among contemporary hunter gatherers and subsistence farmers (2016: 174). Of course copying the practices of others would have been easier after the transition to language.

In sum, warfare among *sapiens* bands may have moderately increased during the late Pleistocene, partly because the demands of their normative cultures heightened us-them tensions with other bands and climactic disruptions led to inter-band and tribal competition for food. But possibilities for migration and prestige-based group transmission also mitigated reasons for making war among these groups. In any case, as will be evident in the next chapter, the scattered raids and ambushes of the late Pleistocene pale in comparison to the ongoing internecine conflicts that erupted throughout the era of chiefdoms and archaic states during and after the transition from hunting and gathering to agriculture beginning about 7.5 kya during the agricultural period.

With the continuing emigration of *H. sapiens* bands out of Africa after 60,000 ya and their spread into Eurasia, the population of our species grew rapidly and it became increasingly important for groups of *sapiens* to distinguish themselves from "outsiders" through language, dress, customs, and other forms of symbolic expression. Within a short time, from an evolutionary point of view, our hunting had decimated and driven to extinction many larger mammals, including—it now appears— the Neanderthals and most other remaining members of the Homo genus. (Whether our species intentionally hunted the Neanderthals as well as intermarrying with some of them remains a debated question).

Sapiens bands grew in size, partly to enable the more efficient slaughter of woolly mammoths and giant sloths, but also to protect the group from other bands, now competing for the same hunting grounds. Eventually, with the decline of big game and the need for a more stable food supply, some hunter-gatherer bands gave up their reverse-dominance coalitions to increase their population size from bands to tribes, gathered around alpha male chiefs who offered to defend them against marauders, and turned to the occasional cultivation of wheat or rice. The agricultural revolution and, with it, the gradual disappearance of most hunter-gatherer cultures, was just around the evolutionary corner. Throughout these dangerous transitional times, conservative tribalism, with its ethnocentrism, religiosity, and sexual (in)tolerance, probably gained more importance for the survival of individuals and their social bands.

3.4 Ritual Interactions in Egalitarian Cultures

Whatever effects egalitarian governance, language-driven norms, and ethnocentrism may have had on warfare, they certainly increased cultural pressure on our species to elaborate more complex religious rituals during the waning years of the Pleistocene. As previously explained, the elevation of everyday actions to iconic, ritual status through mimesis had been available to hominins since the middle years of the Epoch. Building on our *Homo ludens* predispositions, recognized by Gintis and others as a foundational prerequisite for culture, our genus had probably been practicing proto-symbolic play since its beginnings. In his *On the Origins of Stories: Evolution, Cognition, and Fiction*, Brian Boyd links the emergence of rituals and stories to the importance of play in mammalian evolution. Although the evolutionary record is too sparse to substantiate this conclusion, Boyd summarizes the logic and evidence behind it:

> Play evolved through the advantages of flexibility; the amount of play in a species correlates with its flexibility of action. Behaviors like escape and pursuit, attack and defense, and social give-and-take can make life-or-death differences. Creatures with more motivation to practice such behaviors in situations of low urgency can fare better at moments of high urgency. Animals that play repeatedly and exuberantly refine skills, extend repertoires, and sharpen sensitivities. Play therefore has evolved to be highly self-rewarding. Because it is compulsive, animals engage in it again and again, incrementally altering muscle tone and neural wiring, strengthening

and increasing speed in synaptic pathways, improving their capacity and performance. (Boyd 2009: 14)

H. sapiens are the most playful creatures alive. What probably worked evolutionarily for rats, kangaroos, and chimpanzees has continued to work for us.

After 300 kya, the language-ready brain and body of the new species could shape stories and perform rituals with more imagination. In his 2013 book, *Mindvaults*, Radu Bogdan notes that a normally developing four-year old today can build sequences of images and actions in the mind that range far beyond the simple performances of toddlers. These executive abilities come with several new cognitive operations. According to Bogdan, the *sapiens* four-year old can:

> SUSPEND the truth-value and even modal (possibility, probability, etc.) as well asdoxastic (or belief-relevant) status of what is imagined; ADOPT THE PERSPECTIVE of an imaginative projection; use what is imagined from the adopted perspective SUPPOSITIONALLY as a basis or premise for further imaginative projections and inferences; DEPLOY such further projections in a thematically connected manner; [and develop] a METAREPRESENTATIONAL SENSE that one's own offline thoughts represent something imagined (capitalization in quotation). (Bogdan 2013: 140)

While some of these capabilities were probably available to the *Homo* genus before *sapiens* came along, language allowed our species to fully explore and extend them through a range of performances.

Archaeologists have found evidence of ocher, likely used for bodily decoration in performances, at a campsite dated 120 kya, but most evolutionary anthropologists believe that performing probably began much earlier than that, perhaps around 200 kya. Taking on temporary roles for each other and enacting past events around a campfire would not likely have entailed complex production dynamics; initially, *sapiens'* performances may have involved little more than storytelling, accompanied by improvised singing and dancing, perhaps to rhythmic drumming. Probably such campfire performances preceded specifically religious rituals by a 100 thousand years or so. In his *Evolving Brains, Emerging Gods: Early Humans and the Origins of Religion* (2017), E. Fuller Torrey estimates that all of the necessary cognitive capabilities were in place for our ancestors to ritualize their belief in gods by 40,000 ya.

Brian Boyd surmises that *H. sapiens*, already habituated to dancing and singing for other members of their bands, likely believed that their gods would enjoy their performances in the same ways. Boyd acknowledges that we are unlikely to ever find enough evidence to substantiate his own—or anybody's—explanation of the beginnings of religious ritual; he simply presents his ideas as a logical probability, given what we can know about human capabilities, needs, predilections and contexts at the time, despite the lack of empirical facts. Nonetheless, however performing and worshipping came together, the combination has continued to provide a persuasive means of propagating socioreligious practices, norms, and beliefs ever since. Boyd's supposition that rituals grew out of play is supported by linguistic overlaps in several cultures for words that denote both playful and religious activities. The Mbendjele tribe of the BaYaka Pygmies in the Congo Basin of Africa, for example, use the same word, "massana," to mean both ritual and play. Whereas modern *sapiens* living in complex societies today usually set "religion" apart from everyday life, the Mbendjele apparently follow our Pleistocene ancestors in practicing religion as a playful extension of their daily experiences.

Anthropologist Jerome Lewis has explored the different languages used by adult males and females among the Mbendjele and other BaYaka Pygmies and the need for rituals to help male and female groups to cooperate for the good of their bands and tribes. BaYaka languages and rituals are important for our discussion of politics and performances in the Pleistocene because anthropologists believe that they are among the few modes of cultural expression that have survived relatively unchanged from that epoch. For his *Moral Origins: The Evolution of Virtue, Altruism, and Shame* (2012), evolutionary anthropologist Christopher Boehm collated 150 ethnographic studies of surviving "immediate-return" hunter-gatherer societies—meaning that they generally consume what they hunt and gather each day. Boehm calls these groups "Late Pleistocene Appropriates" (LPAs) because he has found significant evidence that many of their cultural practices probably date from the time of *H. heidelbergensis*, roughly half a million years ago. The BaYaka Pygmies are among the more populous immediate-return cultures in the world today. This makes them "hugely significant for anthropology," says Lewis, "since their immediate-return orientation represents such a radically different mode of social organization to the numerous hierarchically organized delayed-return systems that currently dominate human societies" (Lewis 2014: 79). There is little doubt that nearly all Pleistocene hunter-gatherer bands organized themselves as immediate-return societies.

Like many similar groups, the BaYaka Mbendjele practice differently gendered modes of language. The main reason for this, says Lewis, is that BaYaka men and women enculturated each gender to behave differently in the forest. Whereas BaYaka hunters speak softly to each other when they are not listening intently, the women move in large, noisy groups, hoping to scare off the more dangerous animals that live there while they gather yams, berries, and other food. Both men and women tell campfire stories about their times in the forest. When the men act out a successful hunt, they often use their voices to imitate animal calls, a common practice that sometimes lures their prey toward them. In contrast, the women of the tribe tend to celebrate their solidarity as a group in protecting each other from danger and may use their performance time to parody and shame some stupid or abusive acts by men, typically without naming names. Both groups deploy whistling, singing, yodeling, and speaking, and pride themselves on their vocal range and mimetic abilities. Nonetheless, the men and women of the Mbendjele, despite hearing many of the same forest sounds, produce distinctly different patterns of music and speech, some of which are incomprehensible to each other. As Lewis notes, this division prompts both genders of the Mbendjele to meliorate and attempt to resolve their differences through rituals.

Among the most important rituals for the Mbendjele are what Lewis calls their "spirit plays," meant primarily to charm the surrounding forest and its animals into providing food for the Pygmy hunters and gatherers who enact them. Some of the plays ask for specific game animals, while others celebrate the abundance of food for gathering in the forest. Deploying what westerners would understand as both music and language, Mbendjele actors, usually groups of men, women, and occasionally children, engage in singing and dancing for others in the tribe that work somewhat like jazz improvisation, in that the formal organization of the music shapes the timing and variations of each participant. States Lewis: "When listening to the wealth of sound and melody this style produces, it is easy to think that each voice sings randomly, but a sophisticated underlying musical organization constrains and directs innovation and creativity. Each participant's life-long musical apprenticeship has ensured that this musical deep structure is so effectively included that each singer knows how variations can be executed and when to integrate them into the song" (Lewis 2013: 60). Musical expression in the spirit plays is usually a gendered, public activity, allowing collectives of men or women to speak together to the other group. Crucially, says Lewis, "a singing group can say things [in a spirit play] that no individual in the group could say without fearing repercussions.... [This] allows tensions to be expressed and acknowledged even if they cannot be resolved" (Lewis 2014: 88). The spirit plays both enforce the social norms of the Mbendjele tribe and allow festering episodes of shame and hidden guilt to be individually acknowledged.

Understanding themselves and other humans to be one of several societies in nature, the Mbendjele have, says Lewis, "explicitly developed techniques to communicate with many non-human sentient beings around them" (Lewis 2014: 86). The Mbendjele attribute possibilities for spiritual existence and communication to all living things in nature; like most hunter gatherers, their religion is animistic. The tribe uses their musical languages to communicate with their fellow bodies and spirits in the forest. In Mbendjele belief, the forest spirits reciprocate by possessing tribal members, a kind of possession that may float among the singers as different men and women in the tribe pick up and pass on a particular melody or harmonic contrast. Some sounds are recognizable as parts of the language used by the tribe, while others are expressed by animals, insects, and birds in the forest. In J.L. Austin's terms, this makes all of their ritual language directly "performative;" there is no break between "saying" and "doing" in Mbendjele rituals.

Performativity is also evident in Mbendjele initiation rites. The historical conflicts between the genders are on display and open for resolution in the public Mbendjele initiation ceremonies for young women. Preceding this ritual are private meetings in which the female elders of the tribe instruct the new female initiates in matters of courtship, marriage, and reproduction. Based on creation myths that narrate the eventual synthesis of separate male and female tribes, the playful initiation rituals feature performances of war-like battles of the sexes that are usually resolved through laughter and joyful copulation among married couples. They begin with a group of Mbendjele male actors pretending to ambush the women using pig-hunting techniques. The women counter this strategy through a "Ngoku charge," locking arms and rampaging through the camp, threatening to trample the men. The reign of "Ngoku," which is meant to embody an ancient era when women ruled over men, encourages the women to mock male sexual prowess and to openly flaunt their own sexual desires. Descriptions of the Ngoku charge suggest that the action of the initiates moved beyond empathy and high excitement for the young girls and probably many of the older women in the audience to spark emotional contagion. This occurs when a stimulus in the performance triggers the mirror neurons in the brains of many spectators at the same time, leading them to respond together—probably, in the case of the Ngoku charge, with joy.

The Mbendjele initiation ritual for young women works at all four levels of Evan Thompson's understanding of empathetic engagement (Thompson 2007: 382–402). The dramatic coming-of-age ritual of female initiation centered on Mbendjele male hunting and female stampeding depends upon finely tuned "sensorimotor coupling" between the groups to ensure that no one gets hurt in the mock contests. There is little doubt that the women and men are engaging in "imaginary transposition" with one another throughout, seeking to understand the intentions and expectations of their antagonists, both at group and individual levels of interaction. As Lewis acknowledges, these interactions are also a way of acknowledging past embarrassments, slights, and guilts, as each group uses "reiterated empathy" to understand the other. For some participants, these interactions apparently rise to the level of Thompson's "moral perceptions," especially between married couples. Knight and Lewis report that the sexual teasing angers some of the men, but "they usually join in good-humoredly, eventually laughing at their wives' hilarious impersonations of themselves. These reenactments are displayed with such exaggeration and parody as to provoke helpless laughter" (Knight and Lewis 2014: 308). This often allows for a richer expression of sexual affection among married couples after the public ritual concludes. Such playful resolution roughly parallels the Mbendjele origin myth.

If Lewis is correct about the late Pleistocene origins of BaYaka deeply gendered culture and its predominately musical organization, it may be that elements of this mode of ritual signaling in the initiation rites of both young men and women helped to resolve actual conflicts that threatened to tear apart several hominin groups as they struggled toward more egalitarian sexual relations in Africa. The Mbendjele male and female initiation rituals draw from separate, gendered creation myths, which differ in many details. In the male narrative of creation, the men share the sweetness of the honey they have gathered from the forest with the women in order to seize the ritual of sexual reproduction for themselves. The women tell a different story; in their version, they demand one man for each of them to provide her with honey and, in return, they share with the men some of their knowledge about procreation. The narrative thrust of their stories, like the initiation ritual, turns on differences between the force of individual men and the solidarity of the women. Knight and Lewis note that "although Mbendjele men stop well short of endorsing violence against women, they do tend to explore fear and their capacity to inflict physical harm" (Knight and Lewis 2014: 307). Anthropologist Morna Finnegan has suggested that the use of spears in the ritual, which recognizes the potential for rape, remains real for the female participants and underlines why female solidarity "against [male] dominance must be continually reinvented" (Finnegan quoted by Knight and Lewis 2014: 307). Knight and Lewis conclude that the action of the ritual, unlike both origin stories, constructs "a never-ending pendulum swinging between male dominance and its celebratory overturn, between brute force on the one hand and, on the other, female collectivized attractiveness and corresponding power asserted through song, ribald laughter, and erotic play." The crucial point, they add, "is that, finally, men opt to support the women in all this" (2014: 307).

Clearly, this initiation ritual for young women supports governance by prestige and coalition among the Mbendjele, not through alpha male domination. Not surprisingly, coalitions of prestige and persuasion have been and remain the primary mode of governance among the BaYaka Pygmies of the Congo Basin. Before writing *Moral Origins*, Chris Boehm observed in his *Hierarchy in the Forest: The Evolution of Egalitarian Behavior* (1999) that such rituals likely enforced norms of egalitarianism among many hunter-gatherer bands during the late Pleistocene. Although the Mbendjele female initiation ritual advances norms that support different social roles for men and women, these differences do not impinge on the recognition among all tribal members that reproduction carries the same moral weight as male hunting.

The coordinated actions of the men's and the women's chorus in the ritual suggests that both groups learned to expect individuals in their same-sex cohort to sacrifice themselves altruistically, if need be, to save the lives of others in their group. As noted in the last chapter, altruism flourished among *H. sapiens* bands and tribes in the late Pleistocene because the egalitarian ethos of these cultures and the dangers of their surroundings typically encouraged the formation of tightly knit groups of women and men that had to depend upon each other for survival. Like the Mbendjele, many hunter-gatherer cultures encouraged norms of altruism through instruction, practice, and ritual. Together with genetic pressure, the rituals of the BaYaka Pygmies helped to make a lasting difference in our species' predispositions.

Whitehouse's insights into modes of rituals and religiosity brings the prosocial effects of the Mbendjele female initiation into sharp focus. Both dysphoric and euphoric elements are present in this ritual. Negative actions—male aggression with spears threatening to mutilate the women countered by the solidarity of the female Ngoku charge against the men—predominate initially. These actions also include verbal threats from both genders—the threat of physical harm from the men and taunts from the women aimed at humiliating the men. Such dysphoric elements, however, gradually give way to euphoric ones, as singing, dancing, and sexual play take the place of aggression, disgust, and humiliation. The general action of the ritual, for the performers, the initiates, and the audience, moves from dysphoria to a dramatically "comic" resolution in euphoria.

In terms of Whitehouse's two "ritual packages" and their different means of effecting social cohesion, it is clear that this initiation rite primarily deployed identity fusion rather than simple group identification. Recall that fusion occurs when a person's social identity becomes a necessary and important part of his or her personal self-concept. The ritual that officially inducted the young BaYaka women into the tribe as young adults encouraged them to merge their in-group social identity as a member of the female chorus with their sense of themselves. Linking arms with other women and threatening to trample any man who got in their way when they rampaged through the camp embodied this fusion in vigorous ritual action. Whitehouse understands that because "fused" individuals tend to perceive any threat to their in-group as a threat to themselves, they will often respond altruistically in situations when their self-sacrifice could save the group. In effect, the initiation rite prepared the Mbendjele young women to assume more dangerous, even potentially suicidal roles when they accompanied other female adults on foraging missions into the forest.

As noted in Chapter 2, many dysphoric rituals involve what has been called rites of terror. In this regard, the Mbendjele rite of passage into adulthood for females is significantly shorter and milder than many other hunter-gatherer initiations, including the rites endured by Arunta males in central Australia. These Australian hunter gatherers force the young men in their tribes to undergo four dysphoric-fusion initiations into manhood between the ages of 10 and 25. Kidnapped at night, blindfolded, thrown in the air, and taught tribal lore from frightening adult figures, the boys bond together through these terrifying experiences. The second rite culminates in a ritual circumcision, in which the foreskin of each boy's penis is painfully cut off by an adult male using a stone knife. The final rite keeps the young men secluded from the rest of the tribe, frequently silenced, and food- and sleep-deprived for months, as they experience nighttime dances and sacred narratives. In several ordeals by fire, the initiates must repeatedly lie on hot embers, with only a layer of leaves to protect them, and remain there until told by their elders that they may get up. Only at the end of these four initiations were the young men accepted as full members of the tribe. According to Arunta elders, the final ritual "imparts courage and wisdom" and "makes the men more kindly-natured and less apt to quarrel" (cited in Henrich 2016: 162). Nonetheless, some contemporary Arunta males flee the tribe to other parts of Australia to avoid facing the terrors of tribal initiation. This was no doubt an unlikely, perhaps even an impossible option in late Pleistocene times—one of many indications that modern, complex societies have little use for fusion rituals in their mainstream cultures.

Significantly, Arunta young men experience the four initiation rituals of their tribe only once in their lifetimes, but their memory of the occasion stays with them until death. Similarly, Mbendjele women are only initiated once, but they do have the option of participating in Ngoku charges after that. In addition, older Arunta men and Mbendjele women would have had several later occasions to recall their own and other initiation rituals when they helped to plan the rite for subsequent initiates. Before the invention of writing, the best way to imprint the memory of a religious rite was to embed it in a practice that involved high arousal and deep emotions. Pleistocene hunter-gatherer elders were able to impart some information through the stories they told the initiates—a practice that would later accelerate with identification rituals in the Holocene epoch—but the stories, in lieu of literacy, probably varied from generation to generation. According to Whitehouse and Lanman, "Viewed within an evolutionary framework, different societies require higher or lower levels of fusion or identification to fulfill their basic material and economic needs in diverse resource environments... and the two ritual packages evolved, through a process of cultural group selection, to produce the required levels of fusion or identification" (Whitehouse and Lanman 2014: 681).

REFERENCES

Boehm, Christopher. 1999. *Hierarchy in the Forest: The Evolution of Egalitarian Behavior*. Cambridge, MA: Harvard University Press.

Boehm, Christopher. 2012. *Moral Origins: The Evolution, of Virtue, Altruism, and Shame*. New York : Basic Books.

Bogdan, Radu. 2013. *Mindvaults: Sociocultural Grounds for Pretending and Imagining*. Cambridge, MA: MIT Press.

Boyd, Brian. 2009. *On the Origin of Stories: Evolution, Cognition, and Fiction*. Cambridge, MA: Belknap Press.

Deacon, Terrence W. 1997. *The Symbolic Species: The Coevolution of Language and the Brain*. New York: W.W. Norton.

Dor, Daniel. 2014. The Instruction of Imagination: Language and its Evolution as a Communication Technology. In *The Social Origins of Language*, ed. Daniel Dor, Chris Knight, and Jerome Lewis, 105–125. Oxford: Oxford University Press.

Gintis, Herbert, and Carel van Schaik. 2013. Zoon Politicon: The Evolutionary Roots of Human Sociopolitical Systems. In *Cultural Evolution: Society, Technology, Language, and Religion*, ed. Peter J. Richerson and Morten H. Christiansen, 25–44. Cambridge, MA: MIT Press.

Haun, Daniel B., and Harriet Over. 2013. Like Me: A Homophily-Based Account of Human Culture. In *Cultural Evolution: Society, Technology, Language, and Religion*, ed. Peter J. Richerson and Morten H. Christiansen, 75–86. Cambridge, MA: MIT Press.

Henrich, Joseph. 2016. *The Secret of Our Success: How Culture is Driving Human Evolution, Domesticating Our Species, and Making Us Smarter*. Princeton: Princeton University Press.

Hrdy, Sarah B. 2009. *Mothers and Others: The Evolutionary Origins of Mutual Understanding*. Cambridge, MA: Belknap Press.

Knight, Chris. 2014. Language and Symbolic Culture: An Outcome of Hunter-Gatherer Reverse Dominance. In *The Social Origins of Language*, ed. Daniel Dor, Chris Knight, and Jerome Lewis, 228–246. Oxford: Oxford University Press.

Knight, Chris, and Jerome Lewis. 2014. Vocal Deception, Laughter, and the Linguistic Significance of Reverse Dominance. In *The Social Origins of Language*, ed. Daniel Dor, Chris Knight, and Jerome Lewis, 297–314. Oxford: Oxford University Press.

Lamm, Ehud. 2014. Forever United: The Coevolution of Language and Normativity. In *The Social Origins of Language*, ed. Daniel Dor, Chris Knight, and Jerome Lewis, 267–284. Oxford: Oxford University Press.

Lewis, Jerome. 2013. A Cross-Cultural Perspective on the Significance of Music and Dance to Culture and Society: Insight from Ba Yaka Pygmies. In *Language, Music, and the Brain: A Mysterious Relationship*, ed. M. Arbib, 45–66. Cambridge, MA: MIT Press.

———. 2014. BaYaka Pygmy Multi-Modal and Mimetic Communication Traditions. In *The Social Origins of Language*, ed. Daniel Dor, Chris Knight, and Jerome Lewis, 77–91. Oxford: Oxford University Press.

Payne, Keith. 2017. *The Broken Ladder: How Inequality Affects the Way We Think, Live, and Die*. New York: Viking.

Poe, Marshall T. 2011. *A History of Communications: Media and Society from the Evolution of Speech to the Internet*. Cambridge: Cambridge University Press.

Power, Camilla. 2014. The Evolution of Ritual as a Process of Sexual Selection. In *The Social Origins of Language*, ed. Daniel Dor, Chris Knight, and Jerome Lewis, 196–207. Oxford: Oxford University Press.

Sinha, Chris. 2014. Niche Construction and Semiosis: Biocultural and Social Dynamics. In *The Social Origins of Language*, ed. Daniel Dor, Chris Knight, and Jerome Lewis, 31–46. Oxford: Oxford University Press.

Sterelny, Kim. 2018. Adaption Without Insight. In *A Different Kind of Animal: How Culture Transformed Our Species*, ed. Robert Boyd, 135–151. Princeton: Princeton University Press.

Tajfel, Henri. 1981. *Human Groups and Social Categories*. Cambridge: Cambridge University Press.

Thompson, Evan. 2007. *Mind in Life: Biology, Phenomenology, and the Sciences of Mind*. Cambridge, MA: Belknap Press.

Tomasello, Michael. 2008. *The Origins of Human Cooperation*. Cambridge, MA: MIT Press.

Tomasello, Michael. 2016. *A Natural History of Human Morality*, Cambridge, MA: Harvard University Press.

Torrey, E. Fuller. 2017. *Evolving Brains, Emerging Gods: Early Humans and the Origin of Religion*. New York: Columbia University Press.

Tuschman, Avi. 2013. *Our Political Nature: The Evolutionary Origins of What Divides Us*. Amherst, New York: Prometheus Books.

Whitehouse, Harvey, and J. Lanman. 2014. The Ties That Bind Us: Ritual, Fusion, and Identification. *Current Anthropology* 55 (6): 674–695.

Wilson, Edward O. 2005. Introduction. *From So Simple a Beginning: The Four Great Books of Darwin*. New York: W.W. Norton.

CHAPTER 4

Destructive Creation in World History, 9000 BCE–1700

Peter Turchin coined the term "destructive creation" to name the ongoing social transformations that moved our species from small-scale, hunter-gatherer relations to the large-scale, complex relations of nation-states in a globalized world after our transition to agricultural production. He invented the term by inverting economist Joseph Schumpeter's famous phrase "creative destruction," used to describe the dynamics of capitalism. For Schumpeter, "the gale of creative destruction" aptly describes "the process of industrial mutation that incessantly revolutionizes the economic structure from within, incessantly destroying the old one, incessantly creating a new one" (1994: 82–83). Turchin's metaphor, however, describes the incessant historical effects of war, not capitalism: "War is a force of *destructive creation*, a terrible means to a remarkable end" (Turchin 2016: 22). He adds, "And there are good reasons to believe that eventually it will destroy itself and create a world without war" (2016: 22). As Turchin relates in *Ultrasociality*, for the last ten thousand years war and preparations for more war have driven governments to facilitate ever more complex social, economic, political, and communication institutions or risk extinction. This chapter traces the major paths of destructive creation, its significant political effects, and some of its ritual and theatrical performances from the beginnings of agriculture around 9000 BCE to 1700 in the Common Era.

© The Author(s), under exclusive license to Springer Nature Switzerland AG 2021
B. McConachie, *Drama, Politics, and Evolution*, Cognitive Studies in Literature and Performance,
https://doi.org/10.1007/978-3-030-81377-2_4

The forces of destructive creation continued to dominate world history through the transition from agricultural states into the rise of capitalism and up to the end of WWII. In this chapter and the next, I will draw on Jason W. Moore's insight in *Capitalism in the Web of Life* that all economic practices are basically "a way of organizing nature" (Moore 2015:2). For Moore, "the economy" and "the environment" are not foundationally independent from each other because all human systems of economic and social organization, from hunter-gatherer life to post-industrial capitalism, must work through nature to flourish historically. Both agricultural and capitalist modes of production require that our species appropriate what Moore calls "the Four Cheaps" from the natural environment: cheap labor, energy, food, and raw materials. In the agricultural period, appropriated labor power ranged from parents and their children, to slaves and peasants, plus craftsmen and merchants. Agriculturalists appropriated the energy of the sun, soil, and water to produce cheap food, principally crops and animals, for themselves and their masters. And they used such cheap raw materials as iron, wood, and stone to make their tools, build their dwellings and cities, and wage their wars. These "four cheaps" from the agricultural era continued to facilitate economic life even as capitalism began to transform them after 1500.

4.1 From Farming Villages to Macro-States

The gradual transition from hunter-gatherer subsistence to agriculture revolutionized life for most of our species. Instead of moving frequently to find more food, our ancestors settled in one place and transformed their environmental niche to provide more of what they needed for themselves. Local hunting and gathering continued, but with a stable home base, a farming and/or herding community could store food and equipment and gradually upgrade its technology. These advantages allowed agricultural villagers to outbreed the hunter-gatherer bands that occasionally attacked them to steal their food and supplies. It also meant that farming tribes tended to stay in one place, however, even when local conditions changed for the worse. Eventually, agricultural innovations led to improved spades, plows, draught animals, storage containers, irrigation methods, divisions of labor, and larger populations that provided the bases for more complex societies. Throughout the agricultural period, however, tribes of hunting and gathering nomads lived on the periphery of "civilized" cities and empires, mostly on land unsuited for agriculture or grazing.

After the end of the last major ice age, around 11,500 ya, our species began domesticating and harvesting wheat, barley, rye, and millet in the Fertile Crescent stretching from the Mediterranean to the Tigris and Euphrates rivers. Rice followed soon after in China and maize, squash, and beans in Mexico. Agriculturalists domesticated cattle in Mesopotamia and the Indus valley around 8,000 ya and the earliest city-states began soon afterward. Scientists estimate that the Earth's human population stood at about one million before the agricultural revolution. Roughly twelve thousand years later, at the end of the Roman Empire, it had grown to 60 million. Some of the effects of agriculture on human populations were decidedly negative, however. As generations grew to rely on grain-based diets with more starches and sugars, they developed more dental problems, a fact evident in their buried teeth and skulls. Skeletal lesions from leprosy and tuberculous, diseases caused primarily by living close to large human populations and livestock where animal wastes were allowed to accumulate, also increased among ancient agricultural societies. Professor of evolutionary biology Kevin Laland cites studies showing that "Europeans actually became shorter for a time, with males shrinking an average of 7 cm between about 2,300 years ago and 400 years ago...." Despite the flourishing of a small elite at the top of most of these societies, Laland concludes that "agricultural innovations may have repeatedly increased yields, but populations quickly responded by growing in numbers, leaving human misery much as it was before" (2017: 246).

With the transition to agriculture came an increase in warfare. Agriculture led to the cultural invention of private property and private property meant more wealth for some and less for others. Over time, this led to chronic inequality, competition among elites, raiding parties, and organized warfare, gradually undermining and destroying the egalitarian economic basis of hunter-gatherer life. As Turchin explains,

> There is a kind of inevitability about [differentials in wealth], so much so that social scientists came up with a name for it: the Matthew Principle. It goes back to what Jesus Christ said in the Gospel According to Matthew in the New Testament: 'For whosoever hath, to him shall be given, and he shall have more abundance: but whoever hath not, from him shall be

taken away even that he hath.' (Matthew 13:12). In short, the rich get richer, and the poor get poorer. (Turchin 2016: 175)

That is, unless there are accepted institutional practices for redistributing the wealth—practices that evolved and stabilized in some small-scale societies. Among the indigenous peoples of northwestern North America and the Naga living in a mountainous region of northern India, for example, the "Big Men" of their villages fund communal festivals and ceremonies in which they gain influence among their tribesmen by giving away their wealth. By distributing many of their excess goods, they exchange material products for social prestige and dissuade others in the tribe from branding and ridiculing them as hoarders. Potlatches, as northwestern Indians call these ceremonies, are an effective means of preventing the Matthew Principle from overwhelming egalitarian traditions and gradually creating vast disparities of wealth in small societies.

During the agricultural period in areas where warfare became frequent, some of these Big Man societies transitioned into centralized chiefdoms, and from there into archaic states. As J. C. Scott explains, the farming of wheat, corn, rice, and cereal grains made it easier for local strong men to tax nearby peasants: grains are "visible, divisible, accessible, storable, transportable, and rationable" (Scott 2017: 21). Ambitious alpha males could build alliances, win battles, and pursue a military route to power. "Keeping these temporary powers after the war, however, turned out to be an exceedingly difficult proposition," states Turchin (2016: 178). Seeking legitimacy, many upstart chiefs turned themselves into god-kings. According to Turchin, most chiefs sought to combine three main ingredients to transform their initial success into a stable theocratic state: ongoing military victories, ritual leadership to control the religious hierarchy, and "a fanatically loyal retinue" (2016: 178) to quash rebellion from below and prevent potential rivals from seizing power. Religious scholar Robert Bellah also notes that "in a situation of endemic warfare, the successful warrior emanates a sense of mana or charisma, and can use it to establish a following. … It is when the outstanding warrior can mobilize a band of followers that he can challenge the old egalitarianism and, as a successful upstart, free the disposition to dominate from the controls previously placed on it" (Bellah 2011: 2011). Among the most renowned of these ancient charismatic upstarts was Alexander the Great, who transformed a complex chiefdom in Macedonia into a world empire.

The religious cult of Alexander continued to flourish for several centuries after the warrior's death.

While extreme inequality was a regular feature of archaic states throughout the world, human sacrifice in dystopic rituals was less frequent, although it did occur in the early stages of state formation in Egypt and Mesopotamia and became a common practice in northern China, southwest Nigeria, and among the Incan, Aztec, and Mayan states of Central and South America. The elite in most Mayan city-states, for example, used captured soldiers from neighboring warring cities for their sacrifices. They pitted two groups of captives against each other in a ritual, soccer-like ball game, played before the elite and their priests on a formal ball field. The priests apparently used the outcome of this dystopic ritual game to predict the motions of the heavenly bodies, including the sun. In one version of this ritual, to ensure sexual and agricultural fertility for the survival of the elite and their city, the team captain of the losing side was beheaded as a means of appeasing the sun god. For the Mayans, the ballgame was steeped in holy myth. Legend had it that the founding brothers of the Mayan civilization defeated two lords of darkness in this ballgame and then ascended into the heavens to become the sun and the moon. Which side won the game somehow determined the future for the Mayan elite. Competing city-states and their Mayan culture endured in Mesoamerica for almost three thousand years, until, during its long decline, the invading Spanish conquered the culture and its polities in the sixteenth century.

Regarding human sacrifices, Harvey Whitehouse and his coworkers have drawn on Seshat's database to demonstrate that the efficacy of such dystopic sacrifices to legitimize the rule of god-kings is related to the population size and low social complexity of archaic states. As Whitehouse recently explained to science journalist Laura Spinney for an article in *The Atlantic*, human sacrifice becomes a destabilizing force if the population exceeds around 100,000 people: "Our suggestion is that this particularly pernicious form of inequality isn't sustainable as societies get more complex. ... It disappears once they pass certain thresholds" (Whitehouse quoted in Spinney 2018). Whitehouse and Turchin suppose that when some archaic states in the Fertile Crescent began integrating different ethnicities and cultures into their polities in order to grow and defend their empires, terrorizing the inhabitants of a city-state into submission through human sacrifice was no longer a feasible strategy for maintaining power. In fact, it could even backfire, increasing the chances of internal rebellion or external conquest. In the more complex mega-empires that eventually evolved, it also became easier simply to disobey the god-king and evade punishment.

Nonetheless, the success of numerous alpha male god-kings during the long transition from complex chiefdoms to mega-empires—from around 7,000 ya to roughly 2,500 ya in the Middle East, for example—needs to be explained. How could so many of our species have been induced to put aside their moral commitments to equality and fairness acquired during the Pleistocene in order to worship or at least to obey tyrannical rulers? Psychologist Digby Tantum offers some explanations in *The Interbrain: Embodied Connections Versus Common Knowledge* (2018). Tantum draws on Max Weber's notion of charismatic leadership and the distinction between fusion and identification psychological dynamics identified by Whitehouse and others. His primary focus is on the relationship between manipulative leaders and their enthralled followers. Tantum remarks that "worship, veneration, devotion, and adoration," most commonly related "to the intense worlds of fandom or religion," may also afflict the followers of star politicians (Tantum 2018: 228). "Leaders may cultivate such emotions towards their own persons by various means, including ceremonies, public showings, and increasingly sharing 'secrets' about themselves sometimes presented as if they are special privileges for the viewer or listener," states Tantum (2018: 228). He agrees with Max Weber that manipulative charismatic leaders typically inspire moral convictions among their followers by creating narratives of common knowledge shared between them: "Once common knowledge has been established, followers as well as leaders may ensure unswerving adherence to it on pain of exclusion" (231). Tatum adds that this knowledge is often made more convincing by connecting it to master narratives that draw from a common fund of traditional belief. While there is solid evidence that charismatic upstarts deployed these techniques (and others) in the archaic states of the ancient world, their continuing use in successfully binding idolators to demagogues in the fields of politics and entertainment today is also evident.

Turchin does not deny the temporary effectiveness of charisma and hero worship on the politics of archaic states, but, following Weber, questions their staying power. This leads him to disagree with the conclusions of Kent Flannery and Joyce Marcus, whose book *The Creation of Inequality: How Our Prehistoric Ancestors Set the Stage for Monarchy, Slavery, and Empire* (2012), makes the argument suggested by its subtitle. Turchin's *Ultrasociety* reports that elites in Mesopotamia, China, Egypt, and Mexico did gain positions that combined political and religious authority in many archaic states. But he questions whether their monopolization of religion was sufficient to maintain their regimes: "Religion in archaic societies legitimized the pervasive inequalities between commoners and the ruling elite, and clearly this was an important factor restraining commoners from rising up and executing the upstarts. But such restraints were by no means perfect. Peasant rebellions were as much a fact of life in complex, hierarchical societies as peasant deference to their social betters" (Turchin 2016: 145). Sometimes these leaders even used religion to justify their rebellion, as happened in China's Yellow Turban Rebellion (184–205 CE), he notes. Some scholars have argued that god-kings were able to maintain their power in archaic states because they used their divine status to successfully mystify the commoners who worked for them. Turchin, however, suggests that the evidence for the continuing application of terror and coercion, reinforced in some instances by dystopic human sacrifices, is more persuasive. This is partly because "the forager ideas of equality did not disappear. Human beings, including those who live in despotic states, still value fairness and equity" (2016: 136).

Turchin finds evidence for the continuation of Pleistocene beliefs in equality and cooperation in the oppressed peoples of many archaic states, despite their rulers' attempts to erase such hopes through coercion, torture, and occasional human sacrifices. The universal staying power of these beliefs necessarily qualifies the assertions of those historians who would elevate the specific culture of one or another archaic state above the dynamic processes that shaped our species as a whole and that have continued to generate specific sociocultural realities for all cultures at certain stages of their evolution. Seshat has demonstrated that many archaic-state rulers and elites, stuck in the power struggles generated by their coevolutionary level of social complexity, adopted sacrificial state religions that justified such murders. Rather than accepting narrow culturalist explanations for human sacrifices, historians need to understand these developments within a coevolutionary framework.

4.2 Mega-Empires and a Democratic Opponent

The general pattern of destructive creation, by which warfare led to the creation of ever more complex polities, ratcheted up a notch in the sixth century, BCE. In Persia, a mega-empire arose to dominate the ancient Middle East where archaic and macro-states had previously vied for power. The Achaemenid Empire centered in Persia was based on the emergent power of cavalry. Bronze Age chariots had been destructive weapons of war during the archaic-state period, but an armed rider could be more mobile and effective in battle. Although the horse had been domesticated around 3,500 BCE, it took iron smelting, composite bows and iron-tipped arrows, and new technologies of horse control (the bitted bride, the saddle, and stirrups) before armed riders could outmaneuver and overwhelm fully armed infantry. This military technology ensured that cavalry would usually dominate Eurasian warfare from 600 BCE until the development and deployment of gunpowder about two thousand years later.

The new Persian empire (550–330 BCE) stretched about three million square miles at its peak and included some 25–30 million people. It bordered the Indus River in the East, conquered Egypt to the South, and threatened the Greek city-states in the West. It was succeeded by other mega-empires in the region. The Mauryan Empire in India (322–185 BCE) took back Afghanistan and most of Pakistan from the Persians and the Roman Empire (27 BCE–476 CE) dominated the Mediterranean regions where the Persians had ruled and added parts of formerly "barbarian" Europe. Each of these succeeding mega-empires had a population of 50–60 million. A similar dynamic drove up the size of the Han Chinese empire during the same time at the other end of the Asian landmass.

But large populations led to new problems. As Turchin notes, "The unprecedented size of the new [mega-empires] created a burning need for new institutions that would enable these new conglomerates to function reasonably efficiently without disintegrating" (Turchin 2016: 202). Because these empires incorporated hundreds of ethnicities, their populations had little in common. Less than ten percent of those in the Achaemenid Empire spoke Persian, for example, making coordination difficult and helping to foment numerous tribalistic rebellions and civil wars between in-group and out-group populations. Over time, as one mega-empire learned from its predecessor and improvements were made, some stabilization occurred. Successful population management occurred most quickly in China, as the Han emperors moved to normalize the language and customs of its empire, effectively transforming its diverse population into one large ethnocentric group that is now called Han Chinese.

Attempts at such cultural transformations were less successful in the Eastern Mediterranean region, however, where the Persians ran into the expanding military power and vibrant cultural traditions of the Egyptians, Phoenicians, Mycenaeans, and the Greeks that initially blocked their advance. By 550 BCE, the aristocrats who dominated the Greek city-states already controlled the Aegean Sea and were colonizing Sicily, the seacoast of Asia Minor, and extending Greek influence into the Black Sea. The coordination of oar-powered sailing ships with heavily armed infantry on land was the backbone of their military success. Phalanxes of citizen-soldiers called hoplites marching shoulder-to-shoulder and trained in disciplined maneuvers of defense and attack produced city-state cultures that valued military honor and social cohesion. Between 540 and 449 BCE, the Greeks and Persians fought a series of land and naval battles—several of them in mountainous passes and narrow waters that allowed the Greeks to nullify the Persian advantage in cavalry and numbers—that eventually ended in Greek victory and Persian retreat.

Although the city-state (polis) of Sparta had led most of the fighting on land, Athens, the leader of a naval league for the last thirty years of the war, emerged as the dominant power in the Aegean by turning the league of allies into an empire of tribute city-states. In addition to its imperial success, the need for more soldiers and oarsmen during the war years led Athens to grant citizenship to more males. Even before the Persian wars, Athenian politicians had decreased the influence of traditional aristocratic families by shifting more authority to tribal affiliations of citizens, which provided male power for their hoplite phalanxes. By the middle of the fifth century, all Athenian adult males, minus slaves and foreigners, were entitled to vote in the assembly and elect major administrative officials. Though not as politically inclusive and egalitarian as hunter-gatherer bands, which were not economically based on slavery and often featured women in powerful roles, Athens was a democratic outlier among other agricultural polities around the world.

In addition to debates and elections in the assembly, Athenian citizens were also accorded privileged roles in all civic festivals. For the City Dionysia, which occurred in the spring of each year during the fifth century, citizens and young males who would soon reach adulthood honored their warriors who had died in battle, the imperial success of Athens, the god of fertility (Dionysus), and performances of the religious and civic myths that bound Athenians to one another. They celebrated through ritual sacrifices, processions, and choral songs and dances, plus productions of comedies, tragedies, and satyr plays over a six-day period. On the first day of the festival, following sacrifices to Dionysus, the young men of Athens (ephebes) carried a statue of the god to His temple, located below an outdoor theater on a hillside (the theatron) near the Acropolis, the center of religious and political authority. The second day saw dithyramb performances among the ten tribes of Athens, where young boys and men from each tribe competed in singing and dancing choral songs dedicated to the god. On the third day, playwright-directors of five comedies competed by presenting their plays, cast with older men, which ranged among a wide variety of civic concerns and often featured satiric attacks on specific politicians, generals, and philosophers. The fourth day began with a display of tributes from the city-states in the Athenian empire and also included ceremonies honoring the Athenian war dead and crowning current city benefactors. This was followed by the presentation of a trilogy of tragedies and a satyr play, written and staged by the same playwright. (The satyr plays, ribald farces on a subject connected to the three tragedies, featured ephebes wearing erect phalluses and engaging in outrageous sexual activities.) Two more dramatists presented their three tragedies and a satyr play on the fifth and sixth days and the competitions concluded with the judges awarding prizes before evening fell on that last day of the festival. In all, around 2,500 male Athenians participated directly in the City Dionysia, with perhaps another 5,000 citizen-spectators sitting on wooden seats among their fellow tribesmen over the course of the six-day event. (Another thousand auditors from Athens could also find temporary viewing places each day on the hillside above and to the side of the official seating.) The festival centered on a mix of civic and tribal pride, religious observance, mythic history, and the initiation of a cohort of ephebes into full citizenship in Athens.

In his *The Theater of War: What Ancient Greek Tragedies Can Teach Us Today* (2015), theater historian and director-producer Bryan Doerries quotes a statement from the pre-Socratic philosopher Heraclitus that underlines the importance of warfare for ancient Greek theater: "War is the father of all things" (Doerries 2015: 76). Doerries adds, "Many of the greatest humanistic achievements of ancient Athens – arguably one of the most militaristic democracies ever to inhabit the earth – were forged in the crucible of constant military conflict" (2015: 76). This was true not only during the Persian wars for the first half of the fifth-century BCE and but also during the last third of it, when Athens and her allies were fighting other Greek city-states during the Peloponnesian War. The founder of the producing company, Theater of War, Doerries has staged readings of several Greek tragedies for veterans, service members, and their families and found that the plays retain the power "to give warriors of all ranks permission to bear witness to the truth of the experience of war," especially its traumas (2015: 88). In addition to Doerries, I will draw on David Wiles, *Greek Theatre Performance: An Introduction* (2000) in the forthcoming discussion.

While most of Doerries' performances have featured single tragedies by Sophocles, his findings are equally relevant for understanding the one surviving Greek trilogy, *The Oresteia* by Aeschylus, which won first prize for tragedy at its initial performance in Athens in 458 BCE. As will be evident, the action and ethics of the trilogy not only affirmed band of brothers altruism for the ephebes and veteran warriors on stage and in the audience, but also encouraged male Athenians to consider and legitimate important changes in their recent history. As noted, the continuing threat from the Persian empire had led Athens to increase the number of citizens with a direct stake in fighting for their polis, but this expansion of democracy was still fragile. In past times of crisis, some Greek city-states, Athens included, had appointed "tyrants," alpha males with temporary dictatorial powers, to ensure the survival of the polis and many noblemen, the source of traditional power in Athens, still favored this path. Although Aeschylus came from a prominent aristocratic family, he had fought the Persians at Marathon and Salamis—significant Greek victories—and had experienced first-hand what democratic citizenship meant for hoplite solidarity in battle before he wrote *The Oresteia* and entered it into competition in the City Dionysia. As a consequence, his trilogy of plays dramatizes and validates the shift from family-centered to citizen-centered authority in the polis. By the end of *The Oresteia*, hoplite egalitarianism has triumphed over alpha male domination and traditional loyalties.

The first drama of Aeschylus' trilogy, *Agamemnon*, based on Greek mythology, depicts the tragic fall of the aristocratic House of Atreus that ruled in ancient Sparta. As a Chorus of elders awaits the arrival of their King, Agamemnon, from possible victory in the Trojan War, they recall several troubling memories and prophecies that hover over his House. Agamemnon angered his wife by sacrificing their daughter to appease the gods so that he might proceed to Troy and the King's father, Atreus, took revenge on his own brother by tricking him into eating his two sons, whom Atreus had killed and cooked. The Chorus, played by first-year ephebes, march and wheel in phalanx-like formations around the stage, accompanied by an aulos, an oboe-sounding instrument that establishes their rhythm and the mood. The initial choral odes dwell on dysphoric images of the familial murders as the twelve-member Chorus bemoans their possible consequences. Soon after his return, Agamemnon boasts of his triumphs at Troy and shows off his captured concubine, Cassandra, then displays his pride to the Chorus as he enters his palace. Before going inside to join him, Cassandra, doomed to foresee the truth but never be believed, painfully relates the past curses on the House of Atreus and foretells her own death with Agamemnon.

Just after the Chorus wonders if their King must die for the curses on his House, the answer comes from inside the palace as Agamemnon cries out from stab wounds. The Chorus breaks from their phalanx formation and runs distracted as frightened individuals around the stage—an image of shattered social cohesion direct from the battlefield that likely caused remembered fear and panic among the audience of veterans. Immediately after that, the upstage palace doors open to reveal an eccyclema, a low platform on wheels, on which is displayed a sculpture suggesting the agonized death throes of Agamemnon and Cassandra. Behind these bodies draped in gore stands a victorious Clytemnestra, spattered with blood. For the audience, *Agamemnon* climaxed in the Greek equivalent of a dysphoric horror movie. Clytemnestra justifies the murders and threatens to kill anyone who would seek to overthrow the rule of herself and her lover, Aegisthus. Bewildered and afraid, the Chorus worries that only unending bloodshed can come from this act of vengeance. Aegisthus appears, backs up his lover's threats, and reveals that he is the third son of the father whose brother Atreus tricked into eating his first two children—another act of vengeance against the House of Atreus. The Chorus prays openly for the return of Orestes, the son of Agamemnon, to avenge his father's murder, and Aegisthus threatens them for their insolence. But Clytemnestra guides him inside and the Chorus leaves the stage. All that remains is the gore-soaked sculpture on the eccyclema.

The Libation Bearers, the shortest play in the trilogy, begins with the entrance of Orestes and a friend, who move to the down-stage tomb of Agamemnon to mourn his death. They step away when Electra, Orestes' sister, enters with a female Chorus of libation bearers, again played by ephebes and accompanied by an aulos player. It is clear from their chanting and their masks—all stage performers wore close-fitting facial masks—that they have been gouging their cheeks with their fingernails, creating bloody tracks from their fear and lamentation. Given this link to the gore of *Agamemnon* and their subsequent choric odes, the audience is not surprised to learn that these maidens have abandoned marriage and childbearing—their usual roles in ensuring a fertile future for the polis—to act as slaves in attendance to Electra at the tomb of her father. Electra discovers a strand of hair on the tomb, placed there by Orestes, and soon brother and sister are reunited in a common pledge to avenge their father's murder by killing their mother Clytemnestra and Aegisthus. After a short delay, which includes a comic visit from Orestes' old nurse with the Chorus, Orestes confronts Clytemnestra, who curses him before he and his friend take her inside the palace. The doors of the palace open again to reveal a second bloody eccyclema, this time displaying the bodies of Clytemnestra and Aegisthus, with Orestes behind them and his attendants holding the gory robes that had entangled Agamemnon. What initially appears as eye for an eye justice, however, soon falls into confusion and dread. Although the god Apollo had urged his revenge, Orestes feels that his victory is soiled and fears that the familial killings began generations ago may have no end. His final lines indicate that the bloody eyes of the Furies, gods from the underworld, have seized his imagination and they cause him to run from the stage. The Chorus remains to summarize the past murders in the House of Atreus and to agonize about their likely continuation. Though not as dysphoric overall as *Agamemnon*, *The Libation Bearers* has few euphoric moments to lighten its grim pessimism.

The gods that hovered over the actions of the first two tragedies take the stage in *The Eumenides*, the final play of the trilogy, and eventually ensure a euphoric ending. Initially, however, Apollo and the Furies continue the conflict that animated their opposition in the previous play, with the Furies claiming their right to torment Orestes for his matricide and Apollo justifying it because Orestes killed the slayer of his father. In general, the female Furies of the Chorus represented traditional blood ties and aristocratic rights of revenge for the audience, while Apollo stood for youthful energy, patriarchal power, and the revival of tribal loyalties among contemporary Athenians. The ephebe Chorus, dressed in horrible masks and costumes and chanting in unison to the rhythm of the aulos, appears to have the upper hand at first. In Lattimore's translation, they twice repeat this refrain: "Over the beast doomed to the fire / this is the chant, scatter of wits / frenzy and fear, hurting the heart, / song of the Furies / binding brain and blighting blood / in its stringless melody" (Aeschylus 1960: 330–333).

Pallas Athena, daughter of Zeus and patron goddess of Athens, intervenes in the conflict and establishes a court to decide the dispute between Apollo and the Furies. The goddess empowers twelve Athenian citizens to act as jurors and calls the court to order. (This court, the Areopagus, had recently undergone democratic reforms, a transformation that Aeschylus is celebrating in *The Eumemides*.) After arguments from both sides, including testimony from Orestes, Athena announces that, in the case of a tie vote among the twelve jurors, she will decide the case on the side of Apollo and Orestes. When this occurs, Orestes thanks Athena and promises that Sparta shall honor a perpetual military alliance with Athens. But Athena is not finished. Recognizing that the Furies are dissatisfied and threaten to revolt, she promises to transform them from angry avengers into kindly mothers who will watch over Athens and help to guarantee the fertility of its fields, pastoral animals, and women of childbearing age. The former Furies assent and *The Oresteia* ends in reconciliation. Significantly, the mix of ritual and drama in the trilogy has passed from the depths of dysphoric images and actions to an ending that honors euphoric peace between the sexes, roughly the same passage from vengeance and disruption to shared authority and respect that happened in the BaYaka initiation ritual.

To fully understand what the performance of *The Oresteia* meant for the social cohesion of Athens in 458 BCE, it is necessary to place the trilogy within the six-day ritual of the City Dionysia. Whitehouse's criteria for evaluating the sociopolitical implications of rituals can help us to contextualize *The Oresteia's* likely meanings for the ancient Athenians. A significant part of this evaluation must involve the dysphoric and euphoric elements of the ritual as a whole. Part of the reason the overall trilogy could dwell on a predominately dysphoric narrative without leaving male Athenians depressed and anxious was not only because Athens triumphed in the end, but also because so much of the City Dionysia bubbled with euphoria. Dionysus, the god of wine, presided over the festival and wine flowed freely among the citizens throughout the celebrations. Whitehouse also lists feasting, dancing, synchronous movement, and singing as important euphoric ritual elements—all prominently practiced and enjoyed in the sacrifices, dithyrambs, and close-order military maneuvers practiced during the festival. The five comedies performed on the third day of the Dionysia often featured many of these elements, along with raucous laughter. Male sexuality, another opportunity for euphoria, was also on display. As noted, the satyr plays performed after each trilogy during the last three days of the Dionysia often turned the grisly and solemn moods of the tragedies on their heads. *Proteus*, the satyr play performed after *The Oresteia*, is now lost, but its plot was reportedly based on a section of *The Odyssey* in which Menelaus, ship-wreaked on an island with his faithless wife Helen and his male crew, attempts to return to Greece after the Trojan War. It is likely that the combination of a cuckolded husband, a sexy wife, and a chorus of phallus-wielding ephebes presented Aeschylus with numerous opportunities for a sex farce that left the Athenian citizens howling (as a Broadway critic might say) with laughter. Overall, the City Dionysia ritual underlined the honor, legitimacy, and fun of Athenian male culture, their democratic institutions, and their superior military preparedness.

Whitehouse also specifies that ritual analysts look closely at checks on orthodoxy and orthopraxy. Regarding orthodoxy, ritual specialists in every culture typically ensure that the ritual is interpreted correctly. For the City Dionysia, these specialists would have included priests of Dionysus, elders from the ten tribes, the judges for all of the performance contests, and a few officials from the polis. Some of the same people would also have been involved in orthopraxy, monitoring the performances to make sure each part of the ritual conformed to standardized practices. When Aeschylus wrote his trilogy and satyr play and directed his male actors and choruses, he had to work within accepted conventions that limited the length of the dramas, the number of actors and chorus members he could use, and other practices. It is evident from contemporary sources that the predominately phalanx-like movement of the chorus was also standardized, though the degree of flexibility in this convention cannot be known.

In any case, there would have been important social reasons for Aeschylus to conform to the usual musical accompaniment and choreography for his choruses. As Doerries emphasizes, the primary audience for *The Oresteia* were the citizen-hoplites of Athens. What better way to remind these spectators of their previous and continuing solidarity as warriors than to watch Athenian ephebes, hoplites in training, raise their battlefield movements to an art form during one of the greatest festivals of the year. It is probable that the rhythmic chants and movements of the Chorus, supported by aulos accompaniment, excited emotional entrainment in the audience of male citizens. According to musicologist Michael Thaut, entrainment, an extreme form of emotional contagion, emerges from performances involving rhythmic synchronization:

> Several key findings [in this research] have emerged that contribute to an understanding of the neurobiological basis of music and temporal information processing in the brain. Musical rhythm rapidly creates stable and precise internal templates for the temporal organization of motor responses. The motor system is very sensitive to the auditory system. Neural impulses of auditory rhythm project directly onto motor structures. Motor responses become entrained to the timing of rhythmic patterns. (Thaut 2005: 184)

There can be little doubt that entrainment among the ephebes on stage and the adult hoplites in the audience usually inspired altruism in all of them—the willingness to sacrifice yourself for the body next to yours, if necessary, to preserve the phalanx formation and its survival as a fighting unit. Older Athenian citizens needed to socialize the ephebes of the chorus who would soon join them in battle in altruistic feeling and loyalty. Although there is no direct evidence for it, it is likely that some Athenian citizens had a basic, pre-neuroscientific understanding of the links connecting the ritual of the City Dionysia to the military success of their polis. Hoplite altruism, after all, was a significant foundation of success in battle and Athenian democracy.

While many of the adult hoplites in the Athenian audience probably took pride in reliving their battlefield glory when they experienced *The Oresteia*, others would also have recalled their wartime traumas at the festival. Indeed, most probably experienced a mix of these extreme emotional memories, which may have careened between abject terror at the reality of a nearby comrade's death while fighting and an adrenalin rush of strength and joy at the phalanx's success. Contemporary Washington DC theater critic Peter Marks reported on a recent "dramatic collective therapy session" led by Bryan Doerries and involving actor Frances McDormand and others reading scenes from Greek tragedies (Marks 2020: C1). Doerries organized the Zoom performance for more than one thousand English-speaking front-line health workers worldwide who had been assisting COVID-19 patients. What Marks gleaned from the performances and the discussion that followed was "a fresh and moving perspective on how ancient tragedies converse with fresh ones and how bearing witness to suffering remains a profoundly life-changing experience" (2020: C3). Primarily through the agency of the ephebe chorus, many of the fifth-century citizen-hoplites probably experienced a similar collective therapy session by bearing witness to their suffering in the ancient theatron of Athens.

4.3 Accommodating Complexity Through Universal Religions

The Achaemenid Persians survived defeat from the Greeks to flourish for more than another two hundred years, partly by adopting a new strategy to ensure better social cohesion in their armies and far-flung territories. Like other mega-empires, they accepted a universal religion as a means of undercutting divisive tribalism and stabilizing their governance. As disciples, scribes, and patriarchs spread the teachings of Confucius, Zarathustra, Buddha, Jesus, and Mohammed to others beyond their initial ethnic group of origin, several of these religions acquired converts among many ethnicities. According to a recent Seshat study, these religions shared several characteristics: (1) moralizing norms (an emphasis on moral conduct, including the conduct of the ruler); (2) the promotion of prosociality (encouraging acts of charitable giving, for e.g.,); (3) the assumption that gods were more omniscient than rulers; and (4) formalized laws applicable to rulers as well as commoners. By placing potential limitations on the power of political domination and in-group tribalism, these core values moved the political orientations of the universal religions toward prestige, egalitarianism, and xenophilia. In particular, the writings of several universal religions marked an important shift away from the authoritarian norms of alpha male rulers and tribalistic ethnic groups during the previous archaic-state period.

And there is some evidence that these values made a difference in the lives of those living in mega-empires. Zoroastrianism, for example, emphasized all of the four indicators above and its spread played an important role in the Achaemenid Empire. Partly through the influence of Judaism, Christianity also adopted these norms, which in turn shaped its transformation of the Roman Empire in the fourth century, CE. Buddhism, eventually the official state ideology under the Mauryan Empire in India, emphasized prosocial practices and promoted moralizing norms. Like the others, Islam opened a crucial space for thought and action between religious belief and political power. The political implications of Muhammad's preaching were less anti-imperial than those of Buddha or Jesus, but, like Christianity, Islam taught that humanity's first duty was obedience to an all-powerful God, not to an earthly leader. In short, the religious precepts of Buddhism, Zoroastrianism, Christianity, and Islam increased the possibilities for affiliation with members of outgroups and indirectly challenged the supremacy of rulers. Alpha male authoritarians living under these universal faiths could no longer declare themselves god-kings.

Following Whitehouse, it is likely that Zoroastrianism, Buddhism, Christianity, and Islam helped the Achaemenids, Mauryans, Romans, and Muslims to integrate and rule the various ethnicities of their mega-empires. Simply put, these religious traditions could provide a mode of cultural homogeneity among their diverse groups that each empire sorely lacked. Unlike the dysphoric-fusion rituals that centered most earlier religions, these four offered euphoric-identification rituals that allowed their religions to spread rapidly and efficiently. Instead of a few highly arousing rituals that led their followers to recall often painful memories and puzzle out the meanings of their faith over a lifetime, the identification practices of these religions were frequently performed, involved inclusive groups of worshippers, and could be experienced by peoples from all cultures.

Whereas most dysphoric rituals emerged before the spread of manuscript culture, the rituals of the universal religions incorporated euphoric elements that benefitted from the authority of writing. Writing had begun in ancient Sumer, initially as a means of collecting taxes on grain, and continued primarily as an elite activity that furthered the aims of subsequent empires. Used mostly around the palace or the temple, literacy was expensive and time-consuming. For the most part, priests and kings had their scribes write and copy commands that were passed on to illiterate merchants, armies, and peasants through public proclamations. Because literacy favored powerful hierarchies, each of the universal religions allied with the mega-empires of this era established hierarchies to centralize their lines of authority and propagate approved doctrines. Regarding the spread of Christianity in the Roman Empire, for example, manuscript culture helped the early church patriarchs to organize their hierarchy, to determine Christian belief in the holy trinity, and to regularize the ritual of the Christian Mass. Under the Emperor Constantine, Christianity became the official religion of the Empire in 324 CE, a move that ultimately proved more beneficial to the Catholic Church of the European Middle Ages, however, than to the Roman Empire, which fell apart 150 years later.

In India, manuscript culture benefitted the Mauryan Empire, which flourished after the death of Alexander the Great in 323 until 185 BCE.

The Maurya deployed Sanskrit literacy to establish a single system of finance and administration across most of the subcontinent. Following a bloody but successful battle, Emperor Ashoka (272–232) reportedly renounced war and violence to embrace Buddhism. Partly to extend Buddhist teachings, he built a network of roads across much of India and sent missionaries into western Asia, which also helped Ashoka to maintain peaceful relations with the Hellenistic world. Although Buddhism was the official religion of the empire under Ashoka, Buddhist rituals and practices did not extend very far into the population below the elite.

One of the ironic consequences of Ashoka's reign was the apparent beginnings of a revival of Hinduism in some regions, as local Brahmins exerted their influence to resist the spread of Buddhism. Brahmins were at the top of the Indian caste hierarchy, which became normalized around 1,000 BCE. Due to a lack of evidence, the early history of the caste system in India is difficult to piece together and understand, made more so by numerous local and regional variations. Initially, caste seems to have been a Hindu system for classifying people into a hierarchy of groups, with Brahmins (the priestly class) at the top and untouchables (as they were later called) at the bottom. In between were three other classes— (1) rulers and warriors, (2) merchants, artisans, and land-owning farmers, and (3) laborers. In later years, the Indian caste system would expand into a more complex hierarchy involving an elaborate series of social regulations that pervaded many areas of daily life. Whether and to what degree one's caste designation during the reign of Ashoka also mandated who you could marry, what occupation you might take, where you could live, and similar matters cannot be known. What it clear is that the Brahmin class, whose name derives from Brahma, the creator god in Hinduism, was at the pinnacle of the hierarchy. They had reason to resist the spread of Buddhism, especially its egalitarian themes, in order to protect their social status.

Under Brahmin leadership, Hinduism revived and flourished in the gradual disintegration of the Mauryan mega-empire after the reign of Ashoka. In effect, Brahmin priests and scribes lived up to their namesake as the creators of several sacred texts (Vedas) in the years of political fragmentation that followed. They rendered the two great Hindu epics, the *Mahabharata* and the *Ramayana*, which began as oral stories, into Sanskrit manuscripts and also wrote *The Bhagavad Gita*. The *Gita*, which was added to the *Mahabharata*, centers on Lord Krishna and advances the doctrine of the sacred duty (dharma) the devout owe to their class on the caste hierarchy. Brahmins also popularized euphoric cults that encouraged the worship of Krishna, both as a young boy and later as a warrior, and Rama, a benevolent king and family man. In addition, the Brahmins used their Sanskrit Vedas as sources of authority for spiritual practice; in the process, they elaborated and normalized the language for many of the rites of passage celebrated in Hinduism, such as pregnancy, birth, childhood rites, adolescence, weddings, and funerals.

Consequently, when the political disunity ended with the arrival of the Gupta empire (320–550 CE), the Hindu religion was near the center of the cultural life of the empire and dominated what has come to be called the classical age of Indian culture. This included a thriving Sanskrit theater, which borrowed many of its plots from the sacred Hindu Vedas and performed them through poetry and music, complex dances featuring hand and eye gestures, and elaborate, conventionalized costumes and makeup. Specifying exactly how these plays were to be presented was another sacred text, the *Natyasastra*, an encyclopedic work that is one of the longest books on orthopraxy in theater history. The Veda includes the myth that the god Brahma created drama for his people to give them pleasure and instruction. After clarifying the ritualistic practices that must precede a stage performance, the *Natyasastra* discusses two different modes of acting and then enumerates the many components of dance, drama, music, and spectacle that may occur in a Sanskrit performance. These involve many possible movements of the hands and limbs, physical postures, and such details as to how different characters in several dramatic genres should walk across the stage. The book takes up dramatic composition as well, enumerating ten different kinds of plays, rules for their construction, and the chief characteristics of dramatic poetry. The *Natyasastra* is particularly interested in how performers in various kinds of dramatic scenes may spark different emotions (rasas) in the audience; erotic poetry and movement, for example, can inspire love and pathetic scenes can lead spectators to experience sorrow. It also specifies the characteristics of the ideal spectator. Steeped in traditional prestige, the *Natyasastra* continues to inform the decisions of classically trained Indian theater artists when they direct, design, and perform Sanskrit drama.

In addition to drawing on the politics of prestige, Sanskrit theater reinforced the hierarchical principles of the Indian caste system during the Gupta empire. Only high-status male characters—mostly Brahmins and rulers—speak Sanskrit in these plays; the rest (children, women, lower class comics, etc.) spoke in dialects. Also in Sanskrit were the invocation and prologue that began each play and the benedictory prayer that ended it. Significantly, the three extant plays of Kalidasa, the famous playwright and poet of the classical age, used familiar plot devices of romantic love to bend the hard lines of the caste hierarchy toward social cohesion during the Gupta empire. In *Malavika*, for example, a king angers his wife and threatens the caste system by falling in love with an exiled servant girl. Soon after the Queen imprisons her, however, the audience learns that Malavika is actually a well-born princess, clearing the way for her to become the King's mistress. *Vikramorvasiyam* shifts to a different set of star-crossed lovers when a king impregnates a celestial nymph. Although this triggers a curse specifying that the nymph will die when the king sees their child, the curse is lifted, allowing the lovers to remain together. *Shakuntala*, Kalidasa's most popular comedy over the centuries, features a king who meets and marries the adopted daughter of a sage. The king inadvertently angers the sage, who curses him, causing him to forget that he has married Shakuntala until he sees the ring that he left with her. Unfortunately, the ring slips off her finger and is eaten by a fish, but a fisherman returns it to the king and he sees it, breaking the evil spell. Days before Shakuntala is about to give birth to their child, the couple is happily reunited. Similar to the euphoric-identification rituals that helped to spread the popularity of the universal religions, these euphoric comedies ensured that love would triumph over apparent class and gender barriers, eventually allowing sexual passion to work in consort with social hierarchy and cohesion. Although Hinduism was not a universal religion, its reliance on manuscript writing and its extension into dramatic theater helped to legitimate Brahmin prestige and the caste hierarchy that anchored the Gupta empire.

As we have seen, Buddhism, Christianity, Islam, and the other universal religions of the era exerted significant limits on the power of kings, caliphs, and emperors. But they could do little to reign in the depredations of tribalism when distinct ethnicities made war on the basis of religious differences. According to the theologies of Christianity and Islam, God's salvation was available to all humanity. But in order to make oneself worthy of salvation, the individual had to follow certain moral codes and specific religious practices. The unfortunate effect of such restrictions was to transform the faithful of each "universal" religion into separate tribes, divided against each other on the bases of traditions of ritual practice and communities of worship. Within an empire, a single universal religion could become an agent of solidarity, but when in-groups with religious differences collided, the ensuing warfare was more difficult to restrain.

In 1096, a new Catholic Pope declared a Crusade to capture the Holy Land for Christ, beginning a series of wars against non-Christians that were some of the bloodiest contests in history. As world historian J. M. Roberts states, "['The Crusades'] covers a much longer drawn-out and geographically more widely spread series of events than those of the couple of centuries or so which are usually thought of as the crusading era" (2002: 525). In addition to the four crusades fought in the Middle East, Roberts includes the Norman invasion of Sicily and southern Italy, the wars against pagan "barbarians" fought to Christianize Eastern Europe, and the "reconquest" of Spain from Muslim rulers—scare quotes added because most Spanish territory had never been fully Christian before 1492. "The Reconquest was scarcely to be complete," Roberts adds, "before the Spanish would look to the Americas for the battlefield of a new crusade" (527).

4.4 Bunraku Politics in Tokugawa Japan

My final example of politically significant theatrical performances during the agricultural era of world history is from Japan. Although long united as a single country under the rule of an emperor, Japan fell into disunity and civil war between 1467 and 1590. Following several military victories, the Tokugawa family under the leadership of Hideyoshi emerged as the most powerful of the feuding samurai barons. In 1603, Hideyoshi assumed the old title of shogun and used his authority to end the wars, beginning a period in Japanese history known as the "great peace" that lasted until 1868. Political power during the two and a half centuries of the shogunate rested on military might, as the Tokugawa family exercised authority in the name of the emperor but ruled through force and strategic alliances. The shogunate based its legitimacy primarily on a caste system of hierarchical loyalties to social groups and traditions that culminated in loyalty to Japan, symbolized by the emperor. At the top of the hereditary hierarchy were the class of samurai warriors, followed by large landowners and farmers, then artisans and craftsmen. Merchants—because they presumably created nothing but profited from what others had made or grown—were at the bottom. Some groups, including the imperial family and the samurai lords, plus Buddhist and Shinto priests, were thought to be so far above the norms of the social hierarchy that they were not even included in most discussions of it. Other groups, including prostitutes and actors, were excluded from the official ranking because they were simply beneath mention. The shogunate enforced these rigid norms through marital, occupational, and housing restrictions, which were also buttressed by neo-Confucian practices and selective Buddhist and Shinto rituals.

Although the Tokugawa shogunate aimed to create social stability through this caste system, the dynamics of the Japanese economy soon began to undercut these hierarchical relations. Partly as a result of the general peace, agricultural production and internal trade rose rapidly in the 1600s, leading to substantial profits for some merchants and increased wealth in several large cities. Bankers had money to lend and many of the samurai class, now unemployed as warriors, left their fortresses in the countryside and migrated to the growing cities. Despite their higher social status, many samurai became indebted to urban merchants and bankers. Cities were also the locale for print communication during the 1600s, where the samurai, with other literate urbanites, could enjoy woodblock printing and read printed information reliant on the relatively new technology of moveable type. In his *A History of Communications* (2011), Marshall T. Poe emphasizes the importance of print culture for the rise of western Europe into world prominence after 1500, but it also had transformative effects on Tokugawa Japan. By 1700, the bureaucrats of the shogunate were using print to inform the many readers in Japan's largest cities of governmental regulations and printing was pervasive in popular culture.

In particular, the print culture of the 1600s helped to transform the code of bushido, conventionally translated as "the way of the warrior," from a class-based glorification of military duty and practice into a widely shared ethic of loyalty, honor, and virtue. According to social historian Eiko Ikegami's *The Taming of the Samurai* (1995), bushido was a mix of unstable traditions before 1600, primarily passed down through samurai families and manuscript writings, that mostly celebrated bravery and altruistic self-sacrifice in battle. By 1700, printed pamphlets, books, and illustrations, from learned texts to children's books, had transformed bushido into an ethic that all classes could admire and emulate, even though it continued to be identified with the samurai class. Through print media, popular notions of bushido added meditations and tea ceremonies to its celebration of swordsmanship, developed rough exercises for military training into formal regimens of martial arts, and incorporated more of the values of filial piety, loyalty, and restraint from the religious traditions of neo-Confucianism, Shintoism, and Buddhism than had been present before.

In addition to the popularization of bushido, the "great peace" of the Tokugawa shogunate also led to the flourishing of two major types of urban theaters, kabuki and bunraku. Commercial kabuki theaters, several of which began in brothels, sprang up in Edo, Osaka, and Kyota. It did not matter to the samurai lords and their bureaucrats that merchants and craftsmen enjoyed shameless plays and the favors of prostitutes; the ruling class looked down on the classes below them and assumed that such inferior people would always find ways of corrupting themselves. But when lower ranking samurai warriors began to attend the theater, such potential pollution to their class was cause for alarm. In the early years of the shogunate, before the code of bushido had been popularized and widely accepted, the ruling lords of the shogunate forbade their soldiers from attending kabuki theater or even hiring prostitutes where theater was performed. Breeches of such regulations were common, however, and the shogunate struggled to regulate kabuki performances through the early 1650s as a means of discouraging samurai attendance.

Alongside the rise of kabuki theater, a form of puppet theater that was eventually called bunraku also began to flourish in the early 1600s. As in many cultures, hand and string puppets had long been used for dramatic storytelling, but 1610 marks the first record of a puppet play performed by professionals for profit in Japan. Many of the early plays featured miraculous events and moral messages, in an attempt, perhaps, to distinguish them from the kabuki performances that were attracting government censorship. After the establishment of sweeping regulations for kabuki performances at midcentury, kabuki and bunraku artists began to borrow popular themes and plots from each other. Both drew their audiences predominately from merchants, craftsmen, and other male urbanites below the rank of samurai. Less regulated than kabuki, bunraku was popular theater, reflecting many of the interests and beliefs of its spectators. Plays featuring dramatic miracles continued, many linked to the power and beneficence of Buddha, but the themes of love, honor, jealousy, loyalty, and filial piety—themes that were also understood to be a part of the emerging popular understanding of bushido by the 1660s— proliferated as well. Partly because the samurai overlords continued to disdain all theatrical entertainment, there were apparently few orthodoxy checks on the content of the plays. As a result, many bunraku plays dramatized samurai battles and embraced popular bushido ethics.

By the 1680s, with its increasing popularity and evident ability to attract significant artists, bunraku had established stable conventions for theatrical communication, many of which continue to the present day. Derived from narrative norms of oral storytelling, plays written for the bunraku stage depend upon a single narrator or "chanter," who deploys declamation and various kinds of singing to both tell the story and vocally impersonate all of the characters. He sits to the side of the stage, next to a musician, who plays a three-stringed instrument throughout the action. A similar mode of chanting was popular among amateur urban storytellers by 1700. At least two puppeteers support and manipulate each puppet, which today stands about three to four feet tall and is made of wood and cloth; the first controls the head and right arm, the second the torso and left arm. Several puppets are often on stage together, some controlled by a third puppeteer, who manipulates both feet. The puppeteers stand behind a long, narrow, low stage to animate the puppets. By 1700 the manipulators were visible to the audience—most standing with their upper bodies in view of the audience and their faces covered, except for their eyes. For the most part, the audience focuses its attention on the puppets, effectively synthesizing the vocal work of the chanter and the manipulations of the puppeteers. Spectators imaginatively project both voice and movement onto the costume-covered puppet. These conventions for bunraku, based on Japanese folk traditions of storytelling, required no external orthopraxy checks on the conventions of their performance.

Chikamatsu Monzaemon (1653–1725) was one of the most successful playwrights of the early bunraku stage. Descended from a highly ranked samurai family, Chikamatsu received an education in Chinese as well as Japanese classics, but broke from the expectations of his class to perform as a street chanter-storyteller, work backstage in a kabuki theater, and write plays for bunraku and kabuki performers. Although none of his kabuki dramas is extant, nearly 100 of his bunraku plays survive in print editions and several are considered classics of the Japanese literary and theatrical tradition. For several years beginning in 1686, Chikamatsu apprenticed with Takemoto Gidayū, a celebrated chanter and the producer of a bunraku theater in Osaka, where Chikamatsu later became the company playwright. For the last twenty years of his life, Chikamatsu wrote exclusively for bunraku theaters. He was especially praised for his magical-historical epics, five-act plays loosely based on historical incidents that interwove battles and court intrigue with comic and pastoral scenes and divine miracles. Among these puppet plays, most of which took a full day to perform, was his most popular effort, *The Battles of Coxinga*, produced initially in 1715 and frequently revived. Although most of these action-adventure dramas presented at Takemoto's bunraku playhouse played for a few weeks, *Coxinga's* seventeen-month run broke records. And the play has continued popular over the centuries.

A mix of mostly euphoric with some dysphoric elements, *Coxinga's* many performances in 1715 celebrated the major themes and conflicts of popular bushido. Perhaps the major characteristic of the play that the modern reader of Donald Keene's translation will initially note, however, is Chikamatsu's easy mastery of bunraku conventions and his marriage of this form with the content of bushido actions and ethics. The first act of the play begins with the Narrator/Chanter's poetic evocation of a happy and peaceful life at the court of the Chinese Emperor of the Ming Dynasty in Nanking several centuries ago. The Chanter moves from description to dialogue to introduce an envoy from the king of Tartary to the North of China, who requests that the consort of the Emperor, only days away from giving birth to his child, be sent to the Tartar king because he desires her. When this proposal causes consternation in the court, Ri Toten, a General in the Emperor's army, explains that the king's request is based on a deal that he arranged. When China needed rice and millet to avoid the starvation of its peasants, Ri Toten promised the king whatever he desired in exchange for aid from Tartary. Go Sankei, Head of the Emperor's Council of War, objects that the Emperor knew nothing about this arrangement and urges that the Emperor refuse to comply. To which Ri Toten responds, "It is now the task of a loyal minister to sacrifice himself, to calm the emperor, and to wipe out our country's shame. Behold what I do" (Chikamatsu 1951: 103). The Narrator's next line effectively directs the puppeteers manipulating Ri Toten to perform the following actions: "Grasping a knife point downwards, he drove it into his left eyeball, and, placing it on a ceremonial baton, offered it to the envoy" (1951: 103). This action, done in full view of the audience (unlike the blinding of Gloucester in *King Lear*, for e.g.,) could only be performed on a puppet stage and probably had a visceral and shocking effect on the audience. For spectators unfamiliar with the story, Ri Toten's speech about his loyalty to the Emperor and the possible shame of China in its relations with Tartary certainly had the ring of bushido authenticity, but the question of the General's motives would have arisen immediately for the audience.

That question is answered soon, when Go Sankei figures out that Ri Toten's action was the signal for the Tartars to invade Ming China and Toten and his brother quickly kill the Emperor by cutting off his (puppet) head. Go Sankei and his forces slash a path through the invading soldiers to save the consort, who has the heir to the throne in her womb, in order to flee to the coast and escape to Japan. But Tartar musketeers kill the consort and Go Sankei knows that he must perform a Caesarean section in order to save the unborn infant and preserve the possibility of a Ming restoration in the future. After cutting open her dead body and removing the still-living prince, however, Go realizes that the enemy, believing that the heir is alive, will continue to hunt for the prince to murder him. So he takes his own infant son, whom he had rescued in his flight to the coast, kills him, and places the corpse in the womb of the consort. As Go Sankei, the Narrator says, "Admirable child! You have been blessed by fortune. You were born at just the right time and have taken the place of the emperor" (1951: 111). Then Go's manipulators walk him quickly off the stage, carrying the cloth figure of the infant prince. In sum, the narrative of *Coxinga's* thrilling first act dramatizes two major events that are intimately connected to the code of bushido. First, a villainous samurai, Ri Toten, betrays his loyalty to a divinely anointed emperor and kills him. Second, in recompense for this shocking sin, a loyal samurai, Go Sankei, sacrifices his own son to save the rightful heir to the throne in the hope of preserving the Ming empire. Act I of this melodrama establishes the major lines of conflict in *Coxinga* and most of its dominant bushido themes—loyalty to traditional political regimes, the duty to obey superiors, even to the point of familial or self-sacrifice, the virtues of honesty, sincerity, and trustworthiness, and, uniting them all, fealty to hierarchical authority.

Following the mostly dysphoric episodes of Act I, Acts II and III are predominately euphoric and most of the dramatic situations shift from imperial politics to domestic relations within a single family. As Lakoff and Tuschman understand, familial dynamics, including sibling relationships, are often an important source of political orientation for children, shaping their attitudes toward authority and hierarchy. What Lakoff calls the strict father model oriented most samurai families, ensuring patriarchal control over the wife as well as the children and favoring sons over daughters. Banished from the Ming court over twenty years before the action of the play begins, Tri Shiryu, also known as Ikkan, began a new family in a fishing village in southern Japan and raised his only son, Watonai, as a fisherman. Ikkan and his wife live with their son and his young wife, Komutsu, a fisherwoman.

The Narrator introduces these common folk to us at the beginning of Act II and, following the arrival of a strange Chinese woman dressed in courtly attire who floats into their lives on a boat, the twenty-year old son sends for his father. Ikkan recognizes her as Princess Sendan and she recalls him from the Ming court, where Ri Toten arranged his banishment. He tells her right away that "my son has a taste for matters of war and is, as you can see, of a naturally powerful build. ... He will restore the dynasty of the Ming and set at peace the mind of the late emperor in the other world" (Chikamatsu 1951: 119). The action and dialogue of the scene make it apparent that the son is eager to avenge the shame of his father's banishment and will, indeed, do what he can to restore the Ming dynasty. According to the bushido code, restoring family honor can only be achieved through the grace of a new, rightful sovereign. Nor do father and son discuss the merits of the old Emperor's reign. Indeed, it's clear from Act I that life at court was corrupt and the Emperor did very little to defend his dynasty from Tartary aggression. Ikkan's desire to restore the Mings has to do with family honor and traditional loyalties, not the good of the Chinese people.

So Watonai prepares to leave with his father and mother for China while Komutsu, happy to play the obedient wife, agrees to stay behind with the Princess. It is apparent from their relationship that the son respects his aging father's authority and is eager to follow his patriarchal will. While this ready obedience would not have been surprising to the audience, the fact that the fisherman son of a disgraced samurai courtier would have decided to take up arms against a mighty empire with little more than high hopes and a few military ideas certainly added to the fantastical qualities and idealized ethics of popular bushido in Chikamatsu's narrative. Although the playwright drew on published stories for his plot, most spectators in 1715 understood that only family loyalty, resourceful ingenuity, and divine intervention could lead this narrative to a happy ending. But like a film based on Marvel Comics' superheroes today, the story was primarily intended to inspire wish fulfillment, not realistic political expectations.

The climactic scene of Act III demonstrates the centrality of the patriarchal family to Chikamatsu's popularization of the bushido code. After gaining some initial military success in his quest to restore the Mings to the throne, Watonai and his father and mother confront a situation involving familial honor and patriotic shame that can only be overcome through the suicide of Ikkan's Chinese daughter, born to his first wife before he was banished. Watonai, his family, and his soldiers arrive at an impregnable fortress whose lord is allied to the Tartars but also married to Kinshojo, the daughter of Ikkan. Although he has not seen her for many years, Ikkan believes he can convince Kinshojo to persuade her husband, temporarily away from the castle, to form an alliance with Watonai when he returns. She recognizes her aging father standing at the fortress gate, acknowledges her filial duty to her extended family, and promises to try to persuade her husband. Kinshojo also allows her stepmother entrance into the fortress, but asks that she tie herself up in rope as a prisoner so that she will not dishonor her husband, who is under orders to admit no one. Because Kinshojo knows she cannot allow Watonai and Ikkan inside, she also tells him that she will send a signal to them concerning her husband's decision by dying river water that flows under the fortress walls near its gate. Chikamatsu sets up the climax of the act through Kinsojo's dialogue, explaining that she will use White dye if her husband agrees to the alliance and red dye if he does not.

When Kanki, the lord of the fortress, returns, he tells Kinshojo that the Tartar king has promoted him to lead his enormous cavalry into battle. Nonetheless, swayed by his family's traditional loyalty to the Ming, he is also interested in an alliance with Watonai. Kanki recognizes, however, that other Chinese samurai may believe that he has chosen to ally himself and his forces with Watonai because of his wife's relation to him, bringing shame on his family. So he draws a knife and prepares to kill Kinshojo. When her stepmother, still bound with rope, intervenes to protect her, Kanki explains, "If I harm my wife, for whom I feel love and sympathy, it is because I want to ally myself cleanly with Watonai, not influenced by his relationship with my wife, and holding in reverence still the words 'righteousness' and 'fidelity'. … For the sake of your mother's love and your husband's loyalty," Kanki tells his wife, "give up your life" (1951: 137). The strict ethics of Chikamatsu's popular bushido code leave Kinsojo no other choice and she prepares herself for suicide. Recognizing the implications of her death for the honor of Japan, her stepmother states that her suicide "would not only be a disgrace for me, but since everyone would then say that Japanese are cruel-hearted, and give the country that reputation, it would also be a disgrace to Japan" (1951: 138). As Kinshojo prepares to stab herself, she tells her husband, "It would be a disgrace to China if I now held my life so dear that I would not offer it up for my parents and my brother. Now, Kanki, there will surely be no one to slander you and say that your mind was influenced by a woman, so join my father and my brother and add your strength to theirs'" (140). Under the code of bushido, mothers and wives must share responsibility with their warrior husbands for the honor of their nations.

On the puppet stage, Kinshojo's bloody suicide was likely signaled symbolically by a flow of red ribbons from under the fortress wall, leading her brother to believe initially that her choice of red dye meant that Kanki had refused to join his rebellion. But no, Kinshojo had ensured the Japanese–Chinese alliance of honorable samurai families by sacrificing herself for filial and patriotic loyalty, a flow of blood soon after mirrored by the death of her stepmother, who stabs herself in the throat with her stepdaughter's knife. Under the norms and expectations of the Japanese caste system and the popular bushido code of the early 1700s, an altruistic suicide in defense of family honor and loyalty could also ensure the social cohesion of the suicide's social class and even of the entire nation. Although Japanese women of every class had little to gain from bushido ethics and the caste structure, their suicides might be required to redress the dangers of familial, social, and national disgrace. Before killing herself with her stepdaughter's knife, Watonai's mother says, "See Kinshojo! The great desire of my husband and my son has been realized because you have sacrificed your life. ... [The knife you used for your suicide] has altered the destiny of the country" (140). With both of their warrior families now free from shame as a result of Kinshojo's altruistic deed, Lord Kanki can bestow a new name and title on Watonai. From now on, says Kanki, he will be known by his father's title, Tei Seiko, and by "Coxinga," a name meaning "achievement." Watonai, of course, did little to deserve the name in Act III of the play; while he waited outside the fortress, his mother and sister cleared the way for his future success through their suicides.

Act IV continues *Coxinga*'s euphoric march to military success by depicting his links to godly purposes. In Act II, Coxinga had already mounted and ridden a tiger sent against him by his enemies to victory over them and his success in the fourth act continues to harness nature to his divine mission. Two early scenes depict Komutsu, Coxinga's wife, setting off to China with Princess Senden, assisted by a sailor sent by Buddha to magically transport them island-to-island across the East China Sea. The next scene returns the audience to the adventures of Go Sanki, loyal general to the Ming cause, who continues to safeguard the male heir to the Ming throne. In his Journey to the Mountain of Nine Immortals, Go Sanki chances upon two of the immortals playing the military strategy game of Go—the first Ming Emperor and his chief counselor. As he watches them play the game, five years magically pass by, during which Coxinga wins many victories and the heir turns from a boy into a young man. At the end of the scene, the immortals help the prince and Go Sanki—also five years older and now wearing a long beard—to escape to safety from enemy soldiers over a cloud bridge. When the pursuing soldiers try to follow them, the wind blows the bridge away, most of the soldiers fall to their death, and Go Sankei bashes in the head of the general leading them, an ally of Ri Toten, with the Go game board the immortals had been using.

Chikamatsu features the remaining members of Coxinga's family in the climactic battle of Act V, set outside of a fortress defending Nanking. After considering several alternatives for the attack, Coxinga decides on a frontal assault led by his wife, Komutsu, and her soldiers. But his father Ikkan complicates his plans by deciding to end his life through an honorable suicidal duel. When he challenges Ri Toten, the Tartar king orders Ikkan's capture and Coxinga halts his attack when he sees his father strapped to Ri Toten's shield. Go Sanki and Lord Kanki rush in to rescue Ikkan; they bow before the Tartar king, pretend to offer Coxinga's head in exchange for Ikkan's release, and win initial approval from the king for the exchange. But it was only a ruse—the two allies of Coxinga spring up and seize the king, allowing Coxinga to pull his father off of the shield and bind Ri Toten to it in his place. In two final acts of bushido justice, Coxinga lets the king run back to Tartary after 500 lashes but chops off Ri Toten's head and gives to Go Sanki and Lord Kanki the honor of slicing off one of his arms apiece. The villains vanquished, the young Emperor is placed on the throne and the play ends.

In a theatrical production with live actors, it would have been difficult to neutralize the potentially dysphoric effects of several of the physical feats of the last two acts—the fall of enemy soldiers from the cloud bridge into a chasm, the head-bashing of the enemy general with a Go board, and the gory death of Ri Toten. As we know from Punch and Judy shows today (and, for that matter, Roadrunner cartoons), however, the audience effects of inflicting fictitious pain on puppets and cartoon characters can be easily controlled, allowing for some incidents to be empathically excruciating and others to be quickly dismissed. It is clear that Chikamatsu's spectators were not encouraged the feel the pain of Ri Toten and his allies in these and similar episodes. Instead, the defeat of the villains simply chalked up another victory for Coxinga. The consequent social cohesion of this quest is evident in Whitehouse's list of chief examples in his coding for "cohesion" that correspond to several episodes in the play: "honor code," "oath-taking," "enemies of any group members are my enemy," "willingness to die for each other," and "obligations to each others' families." As noted, *Coxinga* legitimated the code of bushido and the caste system it reinforced. Although few of the male spectators enjoying the play were probably samurai, the popularity of *Coxinga* on the bunraku stage suggests that the predominately male, urban audience in 1715 willingly embraced the values and practices of the dominant class in Tokugawa Osaka. There can be little doubt that the play's embrace of bushido and its long-lasting success helped to popularize the militarization of Japanese culture.

This chapter has traced a gradual, winding path through several historical cultures, from the beginnings of agriculture around 9000 BCE to one precapitalist society in 1700, as diverse polities met the threat of warfare from their neighbors by increasing the complexity of their societies in order to protect and advance their interests. Religious rituals also played an important initial role in this transformation, shifting from predominately dysphoric "rites of terror" to rituals that emphasized euphoric elements and consequently legitimated the more complex polities of archaic- and macro-states. Along the way, the ritualized theater of ancient Greece provided a brief democratic alternative to the advances of mega-empires in Persia and elsewhere. The universal religions and their euphoric rituals that predominated between the emergence of Zoroastrianism and the rise of Islam were the first effective checks on the authoritarian power of imperial rulers. This change, plus the rise of print culture, also facilitated the advance of more secular theater in many cultures, including the combination of euphoric bunraku plays and militarism in Japan.

REFERENCES

Aeschylus. 1960. The Eumenides. In *Greek Tragedies*, vol. 3, trans. Richard Lattimore, ed. R. Lattimore and David Grene, 1–41. Chicago: University of Chicago Press.

Bellah, Robert. 2011. *Religion in Human Evolution: From the Paleolithic to the Axial Age*. Cambridge, MA: Harvard University Press.

Chikamatsu, Monzaemon. 1951. The Battles of Coxinga. In *Chikamatsu's Puppet Play, Its Background and Importance*, trans. D. Keene. London: Taylor Foreign Press.

Doerries, Bryan. 2015. *The Theater of War: What Ancient Greek Tragedies Can Teach Us Today*. New York: Random House.

Flannery, Kent, and Joyce Marcus. 2012. *The Creation of Inequality: How Our Prehistoric Ancestors Set the Stage for Monarchy, Slavery, and Empire*. Cambridge, MA: Harvard University Press.

Ikegami, Eiko. 1995. *The Taming of the Samurai*. Cambridge, MA: Harvard University Press.

Lakoff, George. 2008. *The Political Mind: Why You Can't Understand 21st-Century American Politics with an 18th-Century Brain*. New York: Penguin.

Laland, Kevin N. 2017. *Darwin's Unfinished Symphony: How Culture Made the Human Mind*. Princeton: Princeton University Press.

Marks, Peter. 2020. Health Workers Find Solace in Tragedies. *Washington Post*, November 21.

Moore, Jason W. 2015. *Capitalism in the Web of Life: Ecology and Accumulation of Capital*. New York: Verso.

Poe, Marshall T. 2011. *A History of Communications: Media and Society from the Evolution of Speech to the Internet*. New York: Cambridge University Press.

Roberts, J.M. 2002. *The New History of the World*. New York: Oxford University Press.

Schumpeter, Joseph. 1994. *Capitalism, Socialism, and Democracy*. London: Routledge.

Scott, James C. 2017. *Against the Grain: A Deep History of the Earliest States*. New Haven: Yale University Press.

Seshat/Methods/Code Book/Ritual Variables (n.d.). Retrieved: 8 June 2019 from https://www.seshatdatabank.info/methods (n.p.).

Spinney, Laura. 2018. Did Human Sacrifice Help People Form Complex Societies? The Debate Over How Well Ritual Killings Maintained Social Order. In https://www.theatlantic.com/2018/02.

Tantum, Digby. 2018. *The Interbrain: Embodied Connections Versus Common Knowledge*. London: Jessica Kingsley Publishers.

Thaut, M.H. 2005. Rhythm, Human Temporality, and Brain Function. In *Musical Communication*, ed. D. Miell, R. MacDonald, and D.J. Hargreaves. Oxford: Oxford University Press.

Turchin, Peter. 2016. *Ultrasociety: How 10,000 Years of War Made Humans the Greatest Cooperators on Earth*. Chaplin, CT: Beresta Books.

Wiles, David. 2000. *Greek Theatre Performance: An Introduction*. Cambridge: Cambridge University Press.

CHAPTER 5

Capitalism, Tribalism, and Democracy, 1500–1960

As in Chapter 4, I will primarily focus on the universals of our species' political nature and examine their various historical dynamics. The tensions between tribalism and xenophilia, for example, played out differently in the politics of the European enlightenment during the hundred and fifty years before the French Revolution compared to Europe after 1789, when the rise of nationalism led many nations to reject xenophilia and embrace tribalism. Tribalistic nationalism became a major historical force, sparking several wars and leading to the genocides of the past hundred years. Although possibilities for democracy surfaced during the enlightenment, capitalism also brought with it imperialism and hierarchical caste systems based on race that flourished alongside enlightened hopes for more democratic equality. For a closer look at these tensions, I will examine two dramatic performances in the context of their historical cultures, *The Marriage of Figaro* and *The Life of Galileo*. The chapter begins, however, with a consideration of two historical forces that powered early capitalist development, the Protestant Reformation and the printing press.

© The Author(s), under exclusive license to Springer Nature
Switzerland AG 2021
B. McConachie, *Drama, Politics, and Evolution*,
Cognitive Studies in Literature and Performance,
https://doi.org/10.1007/978-3-030-81377-2_5

5.1 Reformation and Enlightenment in Europe

Before the enlightenment could begin to awaken hopes for social and political equality, the Protestant Reformation had to break the hold of the Catholic Church upon the imaginations of educated Europeans. Martin Luther began the Reformation in Germany in 1517 when he challenged the religious and political authority of Roman Catholicism. Not only did Luther rail against Church corruption, but he also insisted that the Bible alone, not the Church hierarchy, was the ultimate source of Christian knowledge and belief. Other reformers had challenged Catholic authority before, but Luther succeeded because he and others used the printing press to broadcast their sermons and demands.

In his *A History of Communications* (2011), Marshall T. Poe asserts that capitalists, bureaucrats, and churchmen pulled print culture into existence (and eventually into prominence) in western Europe after 1500. In contrast, the Tokugawa shogunate prevented enterprising merchants from becoming capitalists and corralled Japanese priests and bureaucrats into using print to propagandize for the regime. In China, despite a longer history of print technology than Europe, there were no emergent economic, political, and religious classes and groups to create the demand for print until the end of the nineteenth century. Consequently, the history of print culture is mostly a western, rather than a fully global history. Print culture also speeded the modernizing advances that gave western European nation-states and capitalists advantages in scientific knowledge, imperialistic power, and commercial dramatic theater.

According to Poe, western print culture had many advantages over the manuscript mode it succeeded. First, it was accessible to more people. The ability to understand legal documents and the Word of God were strong incentives for many and literacy soon spread from wealthy merchants and churchmen to enterprising capitalists and lay religious leaders. By the mid-eighteenth century, readers of newspapers, business documents, and governmental proclamations had vastly expanded the public sphere that first emerged in the Pleistocene in the large cities of Europe, where readers could discuss political affairs, enjoy dramatic plays together, and exchange ideas. Opening governance and business to much wider discussion undermined the old hierarchies and indirectly helped to cause the American and French revolutions. Printing's accessibility also led to standardization in the reach of the law, the rules of the army, and the manufacture of weapons—all of which increased the power of European monarchies and the deadliness of their wars. Although printing was expensive at first, technological changes lowered its price and printing began to be used for entertainment and enjoyment by the 1700s. Given the endurance of printed material over time, practical, scientific, historical, and philosophical periodicals and books could be sent around the world and collected in libraries, adding to human knowledge.

Initially, however, Protestants primarily used the power of the printing press to blow holes in the fortress of Catholicism. By 1530 a massive schism had opened in the Church, as more Protestant sects broke with Rome to begin their own churches and rituals. Turning away from the mostly euphoric rituals of the Catholic Church, the Protestants used dysphoric images in their sermons and prayers to spread conspiracies, anxiety about salvation, and social fears. The dynamic of the Reformation soon churned up dissenters from Martin Luther's beliefs, and these groups, in turn, spawned other opposing sects, as Protestant Christianity splintered into several competing groups. According to historian of witchcraft Kirsten Uszkalo, the anxiety and rage stirred up by the Reformation led to several panics about Satanic possession and witchcraft, which caused the killing of thousands of lower class women. In response to the rapid spread of Protestantism, the Roman Catholic Church mounted a Counter-Reformation between 1545 and 1648 to reform Church practices and revive Catholic faith. Several wars of religion roiled Europe between 1560 and 1648, from civil wars within countries to conflicts involving major Catholic and Protestant powers. The Thirty Years War, begun in 1618, pulled in nearly all of the major armies of Europe. The Peace of Westphalia finally ended the bloodshed in 1648, but not the tribalistic religious animosities the war had aroused. In general, the line separating Protestant northern Europe from the Catholic south in 1650 ran from England through the Netherlands to include Scandinavia and most of the German states in the North, plus the kingdom of Prussia in the East.

While it would be an exaggeration to lay the blame for witchcraft scares and the religious wars at the feet of the printing press, it is certainly true that the print revolution enabled dissenters from Catholicism to mount rival versions of Christianity, which in turn helped to spark nearly a hundred and fifty years of war. By 1650, the tight alliance of monarchical control and religious faith that had maintained political power in the cities, empires, and kingdoms of Europe for four and a half millennia during the manuscript era of communication had been broken. The accessibility, volume, and range of printing had helped to permanently shift these power dynamics in Europe. Most rulers still looked to the established church in their realm—Protestant or Catholic—to bolster their worldly authority, but Catholicism had lost its former monopoly and the exhaustions of war had led some Europeans to recognize that even a grudging toleration of religious differences might be better than ongoing bloodshed.

As historian Siep Stuurman points out, the wars of the period left most enlightenment thinkers eager to "remove the religious dynamite from the body politic" and to redefine the nature of society and politics in secular and scientific terms (Stuurman 2017: 270). The first European to argue forcefully for equal rights based on nature was a seventeenth-century student of René Descartes's, Francois Poulain de la Barre. Following the Cartesian position that all humans were equal in their ability to exercise reason, Poulain asks why social custom nearly always discourages women from reasoning. His understanding of equality influenced subsequent generations of enlightenment thinkers. In Diderot's *Encyclopedia*, perhaps the single greatest achievement of the enlightenment, the entry for "Égalité naturelle," written by Louis de Jaucourt, states, "*Natural* or *moral* equality is that which is based on the constitution of human nature common to all men, who are born, grow up, subsist and die in the same way" (quoted in Stuurman 2017: 280). Unlike Poulain, however, Jaucourt compromises on the rights of women. Although he does not ground male supremacy in nature, he states his belief that husbands should usually have authority over their wives.

In other ways as well, the enlightenment undercut its general commitment to egalitarianism. Stuurman itemizes "four modern discourses of inequality" frequently embedded in enlightenment thought:

> Political economy, a new science that justified social inequality in terms of utility and productivity; biophysical theories of gender, built on the tenet that women were not inferior to men but "naturally" different and complementary to the "opposite" sex; racial classification, a theory that treated humanity as part of the animal kingdom and therefore subject to the taxonomies of natural history. Usually it posited a "natural" hierarchy of races, with white Europeans at the top. The fourth, and in the long run the most consequential..., ordering the multifarious modes of subsistence

and attendant customs found around the globe in a spatiotemporal matrix of more or less "advanced" stages of human development. (2017: 259–260)

Regarding Stuurman's note on "political economy" above, this discourse was extended and complemented by the instrumentalization of nature, which began before the enlightenment in the work of Francis Bacon. As Harari comments in his *Sapiens*, Bacon's *The New Instrument* (1620) argues that "knowledge is power," a position later endorsed by Descartes and others that would energize capitalism as well as enlightenment reasoning. These discourses and their historical effects would impede the full realization of egalitarian rights and out-group equality from the eighteenth-century forward.

The enlightenment would also constrain the full exercise of governance based on prestige rather than alpha male power. Recall that one of the signal developments of our genus during the mid-Pleistocene was the evolution of governance by prestige as a counter to alpha male dominance. Despite the attempts of several women to break into the "all-male club" of enlightened prestige, however, few of them were accepted as equals. Further, even those male thinkers and writers who recognized the intellectual fire power that the addition of women might bring to their causes were constrained by the general sexism of European culture, which continued to privilege primogeniture and male monarchy. The enlightenment shifted cultural prestige away from the Catholic Church and toward a public sphere dominated by *philosophes*, essayists, and scientists, but it stopped short of endorsing prestige for women.

Despite these difficulties, we can see some of the hopes offered by enlightenment natural rights discourse in the eighteenth-century theater of Voltaire, Diderot, and Goldoni. The print revolution boosted professionalism in the European theater, which had attained modest popular and aristocratic success in the market squares and royal courts of Europe by 1600. By 1700, many popular playwrights were publishing their plays and a few royal houses, following the example of King Louis XIV of France, began to subsidize professional theater and opera companies. Voltaire wrote pamphlets, novels, histories, and some of the most popular plays of his time. His *Zaire* (1732) and *Mohammed, or Fanaticism* (1741), by demonstrating the tragic results of not separating religion from the power of the state, indirectly attacked the French king's alliance with Roman Catholicism. Challenging the formal precepts of French neoclassicism in such plays as *The Father of a Family* (1758), Denis Diderot urged the adoption of "middle" genres between comedy and tragedy that could better explore bourgeois family problems through empathy. Like Diderot, Venetian Carlo Goldoni depicted middle-class characters and their distinctive values, which often clashed with traditional aristocratic norms. In brief, the print revolution gave playwrights more prestigious authority in the theater—an authority over theatrical production that would eventually be legalized through comprehensive copyright laws.

Goldoni also helped to transform *opera buffa* (comic opera) from its beginnings in Italian baroque aesthetics into a form that Mozart and others could use to fashion operas that would sustain enlightened insights into political relations lasting to the present day. Professional theaters were open to all, but opera, by the eighteenth century, was primarily intended for aristocratic and royal audiences. As Mitchell Cohen remarks in *The Politics of Opera: A History from Monteverdi to Mozart*, Goldoni's libretti "were often populated by young common women beset by older men or maidservants tussling with desire-filled masters. Opera buffa stirred with everyday life, passions and foibles, licit and illicit, acted out by the kind of figures most viewers might encounter off stage" (Cohen 2017: 274). Goldoni directly influenced Lorenzo Da Ponte, a younger Venetian librettist whose more radical politics forced him to leave Venice for Vienna to avoid arrest. Together, Da Ponte and Mozart would write *The Marriage of Figaro*, *Don Giovanni*, and *Cosi fan tutte*.

5.2 Mozart in Josephean Vienna

Wolfgang Amadeus Mozart arrived in the imperial capital of the Habsburg empire in 1781, where he worked for most of the last ten years of his life as a musician and composer in aristocratic homes and at the Austrian court. Joseph II had inherited the throne in 1780 from his mother, Maria Theresa, whose armies had fought with mixed success against Bavarian, French, and Prussian forces early in her reign to preserve and expand the Habsburg empire. Like his mother, Joseph II understood that Austria must continue to militarize and centralize state power in order to survive the wars and annexations of late eighteenth-century Europe. Accordingly, he reveled in his reputation as an "enlightened despot," a contemporary term of approval for ruling autocrats who sought to improve conditions for their subjects, and pushed reforms designed to get more work from his peasants, better cooperation from provincial noblemen, healthier and literate soldiers, and more taxes through a reorganized and loyal bureaucracy. In short, Joseph II nicely fits Turchin's description of a European monarch driven to add layers of social complexity to the Austrian empire by the "destructive creation" of war and the threat of war.

According to historian Pieter M. Judson's *The Habsburg Empire: A New History*, the lynchpin of these reforms was "the reconceptualization of subjects (*Untertanen*) as citizens (*Staatsbürger*) – that is, as individual men and women with common legal rights and obligations anchored in their unmediated relationship to a central state" (Judson 2016: 51). This transformation of an individual's political status, which took a revolution to accomplish in France, was begun by Joseph II in the 1780s and completed after his death by the next Emperor in 1811 with the passage of the Austrian General Civil Law Code. Judson quotes the text of a 1782 law promulgated by Joseph II, for example, that abolished special regulations against Jews in and around Vienna: "One of [our] principal concerns [is] that all subjects, without distinction of nation and religion... should enjoy a legally guaranteed freedom... to make their living and to contribute by their industry to the general prosperity" (2016: 52). With its attack on traditional rights and hierarchies, this was a radical assertion of the equality of every citizen before the law. "On the whole," notes Judson, "Joseph aimed to give his subjects more opportunities to achieve a better life for themselves because he saw this as the most effective means to increase the overall well-being of the state. At the same time, however, he insisted that he alone was the best arbiter of what was actually good for them" (56). As will be evident, the notion that a well-meaning autocratic reformer became "enlightened" when he instituted laws without the benefit of democratic constraints was a weakness of enlightenment thought that permeated theatrical and political behavior in the eighteenth century.

To explain the political implications of *Figaro* in its time, I will rely initially on Mitchell Cohen's discussion in *The Politics of Opera*. Although some critics have supposed that Da Ponte and Mozart blunted the political points of Beaumarchais' play, upon which the opera was based, Cohen makes a good case that such an argument is plausible "only with an exceedingly narrow definition of politics and ignores variety in the forms – as well as disguises – of power..." (Cohen 2017: 303). Despite their self-censorship, Da Ponte and Mozart's "'comedy through music,' as it was characterized, reinserted politics in the subtleties in the revised libretto, in music, and in dances that speckle it" (2017: 304). Cohen also emphasizes the importance of the law—the guide rails of political possibility put in place by every polity—in the action of *Figaro*. He reminds us that feudal law, contract law, and natural law are all at play in the action of the opera. These legal conflicts involve the ancient right of Count Almaviva to the virginal body of Susanna (Figaro's fiancé), the suit against Figaro by Marcellina, an older woman who claims her right to wed him because of the terms of a contract he made with her and has broken, and what Cohen calls natural law, which finally resolves the plots involving the other two legal conundrums.

Knowing that they were writing *opera buffa* and not an eighteenth-century version of *Law and Order*, Da Ponte and Mozart turned to popular comedy characters and plot devices, many of which Beaumarchais had borrowed for his earlier dramatic version of *Figaro*, and they perfected musical motifs and arrangements that Goldoni had used for comic opera. Mozart's vigorous and playful overture sets the appropriate mood, a term that literary scholar Patrick Hogan takes some time to unpack in his *Literature and Emotion*. Borrowing from cognitive scientist Nico Frijda, Hogan notes that moods in literature typically "render one susceptible to emotion arousal by a large range of events that match the mood's affective tone" (Frijda quoted in Hogan 2018: 39). The droll turns of Mozart's bassoon, the rising tensions of his violins, and the pleasing resolution of his harmonics, in other words, prepare his intended operatic audience for what psychologist and literary critic Keith Oatley has termed "mood-congruent processing" (Oatley quoted in Hogan 2018: 39). Hogan adds that the Sanskrit concept of *rasa* is an appropriate correlate for this notion of mood, because, as its Indian theorists correctly maintain, "*rasa* contextualizes or inflects the more localized emotions that appear in the course of the work" (40). Specifically, Mozart's overture prepares the audience for comedy, laughter, and a happy ending—the generic ingredients of *opera buffa*—even as it also warns of serious conflicts to come.

As noted, Joseph II mounted a campaign to change the patchwork of local and regional laws that governed most of his subjects for uniform laws that could be applied throughout the empire. Among the most important Austrian legal advisors to Joseph II was Karl Anton von Martini, whose magnum opus, published in 1783, emphasized that imperial laws should be based on nature and their logic apparent to every rational person. For Martini, as for many enlightenment thinkers, natural law was a necessary foundation for imperial law; even an emperor, presumably above the conventional laws of the land, must bow to nature. Da Ponte introduces the legal problems of his major characters in the first act and uses the lust of the Count, the foolishness of Marcellina, and the comic vengeance of Bartolo to twist the law against Susanna and Figaro into what seems to be an unresolvable Gordian knot. True to its roots in Jean-Jacques Rousseau's sentimental notion of nature, Da Ponte's plot resolves these difficulties by invoking natural law. It turns out that Marcellina and Bartolo are the true parents of Figaro, who was kidnapped in his youth. Because the marriage of a son to his mother is against nature, Figaro cannot marry Marcellina; in this case, natural law easily trumps contract law. Even more to the point, the natural romantic love of Figaro and Susanna stands in sharp contrast to the designs of those who would enforce unnatural feudal laws to gain villainous ends. The court audience certainly knew about the emperor's interest in legal reform and probably enjoyed Da Ponte's comical treatment of current legal problems.

Of course mere legalities were not about to stop the Count from pursuing and seducing Susanna. When Susanna tells Figaro of the Count's intention toward her and both resolve to stop him in the first scene of the opera, Da Ponte and Mozart use their talents to encourage empathy and sympathy with the loving couple, an emotional tie that continues to grow for most of the opera. The resolution of the legalistic minor plot involving contract law clears away one obstacle, but the steward knows that the only way to relieve the threat to his marriage is to reconcile the Count with his wife, the Countess, who still loves him. As Cohen relates, much of the major plot is resolved through dance and dance-like movements. Early in the first act, after Susanna's exit, Figaro sings the famous "Se vuol ballare," a key to the importance of dance in the rest of the opera:

> If you wish to dance, my little count (*Se vuol ballare, signor contino*)
> If you wish to dance, my little count
> I'll play for you my little guitar
> If he wants to come to my school
> I will teach him the capriole ... (libretto from I, i of *Figaro*, quoted by Cohen 2017: 304).

The threatening circumstances behind "Se voul ballare" set up the conflict between the Count and his steward, Figaro. The music of the first verse suggests a minuet, the stately dance of the aristocracy, while the word "capriole" in the second suggests that Figaro will actually get the Count to move in the bounding motion of a ridiculous goat. Even as the Count is trying to manipulate Figaro and Susanna, Figaro is pulling the Count's strings. "Who will dance to whose tune?" asks Cohen, who concludes, "The opera is organized around this question" (2017: 304). The critic notes several other episodes in the opera when Mozart and Da Ponte use dance to heighten the political stakes of the conflict. As in the movement of the Chorus during the performance of *The Oresteia* in ancient Athens, the combination of dancers moving rhythmically to music likely led to several moments of entrainment for the audience during the performance.

In the finale, set during the night in a pastoral garden (and performed in the eighteenth century under full candle illumination with the actor-singers pretending that they cannot see each other), the dance motif is taken up by Susanna and the Countess, who have agreed to disguise themselves in each other's costumes and roles to trick the Count. The action culminates in the Count professing his love to a woman he believes is Susanna, but is actually his wife dressed in her servant's clothes. Cohen concludes that Figaro—with the help of the Countess and Susanna—has won the dance contest; Almaviva dances to his steward's tune, which includes kneeling before his wife and begging her forgiveness. As Cohen understands, the coauthors drew upon Rousseau's version of natural law for their comic resolution. In *Figaro*, genuine love between an affianced couple is more "natural" than the feudal privilege that grants a master the right to bed a virgin female servant before her wedding night with her husband.

From a contemporary evolutionary point of view, however, it is not at all clear that the resolution of *Figaro*'s major plot has anything to do with natural law. In Debra Lieberman's *Objection: Disgust, Morality, and the Law* (2018), the author points out that the primary tradition of natural law theory depends upon the human emotion of disgust. According to Lieberman, a preeminent scholar on evolution and emotions, disgust is a type of evolved "software program" that regulates human decisions about what to eat, where to touch, and with whom to mate (2018: 10). For this reason, the author points out, nearly all humans are disgusted by the prospect of incest and will avoid mating with close relatives. In performances of *Figaro*, disgust operates in the minds of the characters (and also of the spectators) to drive the steward and his mother, Marcellina, away from the thought of marriage to each other without anyone needing to mention the disgusting Oedipal event they narrowly avoided. Instead, mother, father, and son settle into mutual hugs of relief and sing together about family affection.

But the emotion of disgust is not a part of the love triangle among Figaro, Susanna, and Almaviva, which centers the major plot of *Figaro*. Jealously, anger, lust, aggression, and embarrassment are certainly important in the conflict, and the lyrics, music, and plotting of the opera give glorious voice and action to them. To heighten the conflict, Susanna, though completely loyal to Figaro, is given moments when she is physically attracted to the Count. But the emotion of disgust, which has grounded most versions of natural law throughout history, plays no role in the musical and dramatic romantic triangle. In short, Rousseau's notion of what is "natural" does not fit the major legal tradition of natural law.

Cohen is not wholly to blame for emphasizing natural law in the resolution of the opera, however, because Da Ponte and the court of Joseph II were also confused about what they assumed to be "natural." For Beaumarchais, the originator of the drama's plot, and Da Ponte, as for Rousseau and many intellectuals and rulers, the sentimental side of enlightenment culture led them all to suppose that violating the love between an affianced couple was just as "unnatural" as violating the bond between mother and son. From our perspective today, however, it is clear that the first has to do with culturally relative social norms, while the second, disgust with incest, is indeed a part of our universal nature. Regarding the prospect of copulation between Susanna and the Count, sentimental social norms would indeed work against their sexual union to cause them both some shame, but there is no evolutionary reason motivated by disgust for the Count to turn away from Susanna out of deference to his wife. And on Susanna's part, as Figaro realizes, a servant in her position might secretly violate the social norm to ensure that any children born of her union with the Count would be given a better upbringing that might later benefit her as well.

Consequently, for the happy ending promised by *opera buffa*, Da Ponte knew that he had to characterize Count Almaviva as a man who could be shamed and embarrassed if caught in a compromising position by his wife or even by others of inferior status. Instead of asserting the rights and privileges of his rank on his own estate, the foolish, angry, impetuous, and ultimately sentimental Count is easily put off by Susanna, tricked by Figaro and Cherubino, and finally misled by his wife. At no point in the opera does Almaviva even attempt to beat his wife or take Susanna by force. The Count certainly had the authority to act in this way, but the social norms of Josephean Vienna favored "enlightened" despotism in aristocratic males. Knowing that patriarchal love must triumph in the end, Mozart and Da Ponte could sneak in some egalitarian sentiments and dances, suggesting that stewards might occasionally become the equal of counts in domestic and romantic matters. But it's a tenuous egalitarianism. Even though the embarrassed Count takes the fall for traditional hierarchy in the end, nothing has actually altered the class and gender relations on Almaviva's estate to prevent the same kind of conflict from occurring again, a month later.

Perhaps the one bright spot in *Figaro's* resolution with deep roots in Pleistocene egalitarianism is the legitimation of female solidarity offered by the opera. Susanna and the Countess don't exactly stage a Ngoku charge through the Count's estate, but the effect of their working together to blunt his immoral intentions certainly asserts female power in a moment of danger for both of them. Act IV of the opera demonstrates what two women can do when they cross class lines to foil a lustful rogue and bring him under the control of moderating social norms. While it's appropriate for Cohen to give some credit to Figaro for turning the Count's amorous intentions toward the Countess, it's the cooperating women, not he, who finally gets the master to dance to a different tune. In fact, Figaro—in the dark about his wife's plan to trade clothes and pretend to be the Countess—is useless in Act IV until very near the end of the opera, consumed as he is by jealousy, fear, and anger at Susanna, who he imagines is about to betray him with the Count.

Figaro's 1786 premiere and subsequent successes across Europe clearly place the opera on the far side of euphoric with regard to Whitehouse's euphoric-dysphoric spectrum. Recall that the contrast between dysphoric and euphoric rituals hinges on events and emotions that encompass such polar opposites as: risk of death/feasting, mutilation/dancing, plus disgust/mirth and fear/curiosity. *Opera buffa*, of course, is supposed to celebrate singing and dancing, two other activities typically featured in euphoric performances, and *Figaro* easily lives up to its generic expectations. Even the minor characters Marcellina and Bartolo, who might have been depicted with a touch of dysphoric grotesquerie, given their initial obsessions, are graced by Mozart and Da Ponte with traits that eventually inspire normative affection because they must transform from cranky opponents of our hero and heroine into natural symbols of familial love.

As Whitehouse predicts, the euphoric emotional tone of the opera relates to its embrace of social cohesion. A sentimental version of Rousseau's understanding of natural law that also endorses enlightened despotism, *Figaro* rewards sincere affection and conventional social norms over male lust and power. Cohen adds that the finale, though "[hardly] a call for revolution" (315), does end with reconciliations across class and gender lines. True, but *Figaro's* narrow focus on resolving the immediate amorous, social, and legal complications of its major characters excludes consideration of the larger structural dynamics of the *ancien regime* where class, gender, and sex are concerned.

5.3 Capitalism and Imperialism, 1500–1900

According to economic historian Jason Moore, the Cartesian revolution shifted ontological status from relationships to entities; matter, people, and ideas became discrete things rather than interactive processes. In so doing, the enlightenment "strongly favored the idea of a purposive control over nature through applied science" (Moore 2015: 20). As a result, says Moore, "The dualism of Nature/Society is complicit in the violence of modernity at its core.… For this dualism drips with blood and dirt, from its sixteenth-century origins to capitalism in its twilight…" (2015: 4). Descartes had learned to think about technology and nature in these ways by living and working in the Dutch Republic, the first world power to control a capitalist empire. While there in the 1630 s and '40s, Descartes applauded the Dutch reorganization of nature in the early capitalist period through better sailing ships, the technology of print, standard contracts, a stock exchange, cartography, telescopes, and other instruments of power that the Dutch were either inventing or implementing.

Before the Dutch could instrumentalize nature in these ways, the early capitalists in the Netherlands and elsewhere benefitted from the European wars of religious conflict, dynastic succession, and imperial conquest between 1480 and 1620. The German Peasants War in 1525, for example, though sparked by the Reformation, quickly led to a class struggle in which the peasants rebelled against the restoration of feudal relations and taxes in the wake of a mining and metallurgy boom in Central Europe. As before in world history, wars of destructive creation caused European powers in the early modern period to add layers of complexity to their rule in order to protect their regimes and advance their interests. The logic of political survival led European polities to search for untapped frontiers of resources and revenues, whether in the Vistula valley of Poland or on the Atlantic coast of Brazil. Consequently, the long sixteenth century saw the exhaustion of Mediterranean forests and shifts in forest product appropriation from Poland-Lithuania to Norway and eventually to North America, the extraction of cheap domestic peat for energy in the Netherlands, and the gradual rise in English coal production after 1530. One important result of early capitalism was a change in European diets, as the "Columbian exchange" moved animals and crops from the New World to the Old and diseases from the Old to the New. (See Alfred Crosby, *The Columbian Exchange: Biological and Cultural Consequences of 1492*). This process encouraged European aristocrats and entrepreneurs to eat more meat, feed the newly available potatoes and corn to their peasants and workers, and allow smallpox, measles, and influenza to decimate American native populations.

The capitalism-imperialism link also led to the development of the first sugar–slave nexus in Madeira and a profitable world sugar economy in Brazil after 1570. Despite the belief that all humans were equal before God, there were limits to Christian equality, shockingly evident in the enslavement of indigenous new world tribes and Africans in the Americas. As several historians and anthropologists have pointed out, the invention of racially based slavery was new in world history. According to anthropologist Ashley Montagu, Europeans invented "an arbitrary and superficial selection of traits" to sustain their new caste systems. "The idea of race," he wrote, "was, in fact, the deliberate creation of an exploiting class seeking to maintain and defend its privileges against what was profitably regarded as an inferior caste" (quoted in Wilkerson 2020: 66). The embrace of racially based slavery tipped post-1500 European imperialism toward tribalism, hierarchy, and domination, a nexus of power relations that would continue to divide and entangle humanity to the present day. When enlightenment reason and political revolutions made egalitarianism—that commonplace assumption of late Pleistocene hunter gatherers—again thinkable for humanity, it was reserved for "White" males of European ancestry.

African slavery and new world raw materials helped capitalism to reorganize the "four cheaps" of nature, especially Cheap Labor and Cheap Resources. For Jason Moore, what was new about capitalism was that appropriation must stay ahead of the expanding domain of commodification (including worker exploitation) in terms of total costs to enable the system to continue to renew itself. As Moore notes, this insight changes our usual thinking about capitalist development. Although capitalism does expand commodification in the search for profits, more importantly it also "expands to shift the balance of world accumulation towards appropriation" (Moore 2015: 102). The British- and American-led industrial revolutions, for example, unfolded through epoch-making appropriations of the accumulated work/energy of fossil fuel formation (coal, then oil) and the accumulated work/energy of humans raised to adulthood outside of the commodity system" (2015: 102)—dispossessed peasants and slaves. This leads Moore to emphasize the importance of new frontiers for capitalist renewal and the inevitable linkage between capitalism and imperialism. This dynamic was true as well for Dutch hegemony in the seventeenth century, which depended upon cheap timber for shipbuilding after 1570. In addition, notes Moore, the Dutch "imposed a new colonial regime between the 1650s and 1670s [in what is now Indonesia], securing a monopoly over the clove trade [in part] through... the large-scale relocation of indigenous populations from the interior into new colonial administrative units suitable for labor drafts" (185). The Dutch also dominated early trading in African slaves.

Great Britain's combination of industrial and imperial power during the long nineteenth century established it as the next capitalist world hegemon from 1840 until 1914. Powered by coal and built on printing presses, railroads, and steamships, "its crowning achievement," says Moore, "was a revolutionary advance in appropriation, as capital's steel tentacles penetrated far flung peasant formations, from South Asia to Eastern Europe, setting free vast rivers of Cheap Labor power" (Moore 2015: 153). This also included US slavery after 1700, which eventually supplied the cotton mills of Manchester, later to be fed ginned cotton by other Cheap Labor in Egypt and India. The 1840 date at the beginning of British world hegemony marks the start of the First Opium War, when British warships blasted Chinese coastal towns, slaughtering thousands of civilians, so that British cartels could continue to sell opium to the Chinese. By 1900, the sun never set on British coaling stations (on land as well as sea) across the world.

The world fairs that began with the Crystal Palace Exhibition in London in 1851 celebrated the many links connecting capitalism and imperialism between 1850 and the beginning of the Great War in 1914. Although begun to glorify the progress and superiority of their nation-states, world fairs soon settled into promoting national empires. While the Crystal Palace had included exhibits from selected parts of the British Empire, the Paris Exposition of 1855 featured representations from every French imperial possession in the world. After the Germans annexed two French provinces in the wake of the Franco-Prussian War (1870–1871), the French increasingly turned to imperial glory abroad and expositions at home to emphasize their greatness.

The Paris Exposition of 1889 was the first of many world fairs to exhibit native peoples to the gaze of White visitors. The US followed the trend in 1893 at the Chicago Columbian Exhibition, where seventeen tribes and villagers from parts of the British and French empires were housed in their native settings near the Midway Plaisance, an area of the fair that also featured "freak" shows and other carnival acts. At the Pan-American Exposition in Buffalo in 1901, the organizers exhibited Native- and African Americans along with other "primitive" peoples of the world. Most of the fairgoers apparently enjoyed the spectacle of Indians performing war dances in traditional attire and African Americans hired to portray happy antebellum slaves in a popular exhibit called "The Old Plantation."

In retrospect, the world fairs celebrated an enlightened Cartesian perspective that elevated western civilization and progress above mere nature, which could only be useful if appropriated for capitalist profit and imperial glory. In the terms established by Whitehouse, the narratives of imperial progress boosting the fairs were exercises in euphoria; they parroted back to their spectators practices and beliefs about capitalism and imperialism that were broadly appealing to western audiences. In addition to justifying vast material inequalities based on tribalistic racial distinctions, the fairs elevated their own national and imperial ethnicities above the inferior out-groups from the colonies. World fairs were also exercises in historical amnesia, intentionally hiding the alpha male power grabs that began most imperial takeovers and substituting a mask of paternalistic concern that kept the largely native bureaucracies of capitalist appropriation ticking along under imperial rule.

5.4 European and American Tribalisms, 1770–1945

Although the period of European and American political revolutions, begun in 1776 and ending in 1848, usually marched under the banners of liberty and equality for all citizens, most of the revolutions did more to excite the tribalistic passions of nationalism than to deliver the structural reforms that could guarantee democratic inclusions and procedures. The French Revolution and the Napoleonic wars that followed enflamed tribalism throughout Europe. While the French believed they fought to defend their revolution, the monarchs of Europe soon learned that the best way to rouse their countrymen against Napoleon's armies was a nationalistic appeal to their own people. Although many nationalists professed some interest in the enlightenment ideals of individual liberty, private property, and constitutionalism, they did not always honor them in times of political crisis and warfare. Instead, most endorsed a version of cultural nationalism, the belief that their tribal cultural traditions were unique and deserving of a strong state to glorify them. Racial nationalists went a step further, mixing traditional notions of racial superiority with cultural nationalism to produce a hierarchical and authoritarian brew that was particularly potent in Germany, Brazil, the US, and Japan. Most German racial nationalists believed that Germans were inherently superior to Jews, Slavs, and Frenchmen, racial nationalists in the US and Brazil elevated an invented "White race" over those descendants living in their nations that had previously been enslaved, and many Japanese believed themselves superior to Whites and other Asians.

Whatever the type of nationalism embraced, the idea that a people, a nation, loosely united by a common language and/or culture, has an inherent right to its own geographical and political state would have seemed absurd to most of humanity before 1750. Most people in the world owed allegiance to distant rulers and/or local chiefs and their cultural lives were bounded by topographical, spoken, and regional differences. As historian Benedict Anderson notes, the fellow feelings that supposedly undergird nationalism have to be invented and continuously reaffirmed because a nation-state is actually a gathering of strangers. The national people of such a polity is consequently an "imagined community" (Anderson 1991: 6) in need of constant affirmation. Along with national flags, anthems, and other performances, print communication, especially through textbook education and national-language newspapers, was the primary means to encourage citizens to imagine themselves as loyal members of a nation-state. By 1914, several wars for national unification had rearranged the maps of Europe and the Americas to reflect what the elites of each country and an increasing number of "the people" understood as their nation.

Although nationalism remained the dominant mode of tribalism across the capitalistic world through 1945, the industrial revolution's reliance on coal gave European democrats and socialists the opportunity to challenge nationalism's reign by expanding working-class solidarity. In *Carbon Democracy: Political Power in the Age of Oil* (2011), historian Timothy Mitchell examines the connections between capitalist revolutions in cheap energy and the growth of egalitarian democracy. Noting that initially there were very few sites in Great Britain, France, Belgium, and the US where abundant quantities of coal could be mined in the late nineteenth century, he explains that this relative scarcity gave some workers key political advantages in their nations. Because of the insatiable demand for coal and the relative ease with which miners, railroad men, and dock workers could stop its transportation to factories and electrical generating stations, these workers were able to join together in labor unions and, eventually, to enforce their will in democratically elected political parties. States Mitchell,

> The flow and concentration of energy made it possible to connect the demands of miners to those of others, and to give their arguments a technical force that could not easily be ignored. Strikes became effective not because of mining's isolation, but on the contrary because of the flows of carbon that connected chambers beneath the ground to every factory, office, home, or means of transportation that depended on steam or electric power. (Mitchell 2011: 21)

In effect, labor control of the "flows of carbon" swelled the ranks of working-class protest in the industrializing West and eventually powered democratic reforms.

For most of the nineteenth century, conservative elites and emergent capitalists limited the vote in Europe to property owners. Although Jacksonian democracy expanded suffrage to nearly all White males in the US, other democratic reforms—such as free and fair elections, the protection of minority rights, social insurance, plus welfare and health care—were not offered in the US or most European countries until the twentieth century. In Great Britain, the first nation-state to industrialize, craft workers (carpenters, plumbers, masons, etc.) began unionizing in the 1850s, followed in the 1880s by industrial and unskilled workers. The Great London Dock strike of 1889, which stopped the transportation of coal and other products by water for forty-one days, was followed by the formation of the Transport Workers Union. The gradual success of strikes and further unionization led to the formation of a small Labor Party, which acted in concert with the Liberal Party to gain parliamentary power in 1906. The Liberals and Labor passed laws enacting accident and old-age insurance and a moderate minimum wage bill, plus the removal of some restrictions on strikes and other union activities. In the UK, stopping the flow of coal led directly to democratic reforms in a few short years.

Compared to the roughly two million union workers in Great Britain in 1900, only 250,000 workers had unionized in France and 850,000 in Germany by that date. Their paths to popular democratic reforms were consequently more circuitous. By the 1890s in both countries, varieties of socialism flourished more easily than in the UK. In France, the socialists joined with other parties in the Third Republic to pass anticlerical legislation funding public schools. They also sided with other Republicans and radicals during the tempestuous Dreyfus Affair in the late 1890s, pushing for reforms that helped to curb the strength of the monarchists and the army in French politics. Having successfully united the German Empire in 1871, Chancellor Bismarck sought to limit democracy in the Reichstag by maintaining the absolutist power of the emperor and his chancellor to govern the nation. Nonetheless, Bismarck worried that the growing power of the socialists would destabilize the country and sought to cement working-class loyalty to the Reich by passing an extensive program of social legislation that insured workers against accidents, sickness, and old age. Despite the Chancellor's attempt to kill socialism with the most generous state insurance package in Europe, the socialists elected more delegates to the Reichstag in 1890 than they had in 1878.

Of the major industrialized nations in the nineteenth century, the US was the least responsive to working-class demands for more democracy before the Great War. US law did not recognize workers' rights to form unions, bargain collectively, or strike until 1935. The result was a series of bloody confrontations after 1870 between capital and labor in which management was free to use spies, private police, strikebreakers, and violence to curb working-class demands. Another impediment to unionizing workers in the US was White American racism against immigrant populations as well as African Americans. US capitalists systematically deployed Black citizens and various ethnic groups against White workers whose relatives had emigrated earlier. In addition, there was a political obstacle that distinguished US labor history and democratic progress from its European counterparts—the US constitution. Because the Founding Fathers did not foresee the emergence of political parties, they enacted a Constitution that had the effect of limiting governance to two major political parties at the national level. Whereas parliamentary systems are inherently open to several parties and coalitions among them may govern successfully for a long time, the same has never been true in the US. Labor and socialist parties in Great Britain, France, and Germany started small, but joined coalitions, gained legitimacy, boosted their numbers in subsequent elections, and emerged with a substantial measure of political power by 1914. In Germany, for instance, the Social Democrats were the most numerous party in the Reichstag in 1914, despite Bismarck's earlier attempt to kill it. In the US, Republicans and Democrats occasionally courted working-class voters, but workers could not form their own party and gain success in Congress (much less the Presidency) without challenging the two-party system. Coalition-building across class lines—an egalitarian practice dating from our species' time in the Pleistocene—was mostly denied to US workers until the New Deal and seriously impeded working-class political effectiveness.

US working-class entanglements with racism and immigration turned many American dramatists away from industrial labor problems as legitimate subjects for commercial drama. In contrast, the greater prestige of European working-class causes led prominent European dramatists to write several plays between the 1880s and 1920s with working-class heroes and victims. In France, Émile Zola and Henri Becque wrote socialist plays featuring downtrodden peasants (Zola's *The Earth*) and scheming bourgeoisie (Becque's *La Parisienne*). While living in Paris, Swedish playwright August Strindberg penned *Miss Julie*, about a decadent aristocrat seduced by a male servant, and saw it produced in an avant-garde theater. German socialist playwright Gerhardt Hauptman wrote *The Weavers,* a play with a group protagonist about the exploitation and rioting of Silesian weavers in the 1840s. George Bernard Shaw emerged as the most outspoken socialist in Great Britain. A champion of ending capitalist oppression through gradualist means, Shaw's plays attacked slum-landlordism, capitalist profits from prostitution, imperialism in Ireland, and the idiocy of armies and war. In 1905, through *Major Barbara*, Shaw's audience learned that social conscience without economic power is useless and, finally, immoral. Apart from Shaw's comedies, which bristled with euphoric wit, most other socialist plays chiefly dramatized the dysphoric elements of painful and unjust situations;

they typically featured fear, disgust, and humiliation. By moving from grim oppression to the joyful rioting of workers in the finale, however, *The Weavers* managed a rough balance of both elements, even though, historically, their brief rebellion was ruthlessly put down.

Despite the relative success of working-class theaters and politics in Europe, the outbreak of the Great War in 1914 demonstrated that the nascent egalitarianism that might have produced international solidarity among workers was no match for the various tribalisms of European nationalism that quickly put workers in uniform to fight each other in defense of their nations. Many Socialists had hoped that class solidarity might trump nationalism—and some of their leaders were jailed for pacifism during the war—but no general strike occurred. The Great War also fractured the emergent cosmopolitanism of Western bourgeois culture, which had grown more intertwined since the 1850s and shared many cultural affinities by 1914. Besides the length and carnage of the war, the biggest surprise for the bourgeoisie was the eruption and success of the Russian Revolution, which put anti-capitalist Bolsheviks in control of a major European empire and nearly caused the political collapse of Germany after the war. Although Britain, France, and the US sent troops into Russia to fight against the Reds in the civil war that followed the revolution, their intervention did not stop the formation of the new Soviet Union.

Nonetheless, the success of the revolution and the end of the war did little to challenge the ideologies and passions of cultural domination and racial nationalism. Indeed, US President Wilson's "14 Points," which provided part of the context for the Treaty of Versailles that ended the war in 1918, validated the desires of many cultural and racial nationalists by embracing the principle of cultural self-determination. This initiative, advanced by a President who favored racist policies in the US, justified the creation of a state for every group of self-defined national people, an impossible expectation given the scattered population patterns of American and European settlement. The Great War broke up the defeated Austro-Hungarian and Ottoman empires and altered the western border of the new Soviet Union to create fourteen new nation-states in Europe by 1923, but few emerged without nationalist bloodshed. In his *The Vanquished: Why the First World War Failed to End* (2016), historian Robert Gerwarth details twenty-seven violent conflicts between 1917 and 1920, most of them due to tensions among rival cultural or racially defined groups. German nationalists killed Latvians,

Greeks attacked Turks, and several nationalities in the former Austro-Hungarian Empire mounted anti-Semitic pogroms to kill Jews. The Turkish campaign of genocide against the Armenians might have provided the sharpest warning of the horrors to come, but few Europeans were paying attention. It would take the revelations of the Holocaust, which occurred for most westerners after the end of WW II, to awaken the world to the tragedy of racist nationalism and its brand of authoritarianism.

Before the next world war, however, two new audiovisual media emerged that would help to fan the flames of modern tribalism and domination, radio and motion pictures. By the end of the nineteenth century, as media historian Friedrich Kittler affirms, the telephone and phonography, invented in 1876 and 1877, were separating sound from the materiality of the human body for millions of listeners. These audiophonic inventions carried the intimacy and immediacy of the human voice on invisible sound waves that lacked the apparent concreteness of live performances and still photography. In the past, "hearing voices" had been a sign of religious possession or mental instability and these traditional attributes clung to the affects of the new audiophonic media. As Kittler points out, these media led to innovations in musical composition for some, affirmed spirituality for others, and increased tribalistic bonds for many. The rapid rise of radio broadcasting after 1920 spread these effects to national publics. On popular radio broadcasts in the US, for example, listeners willingly suspended their disbelief to accept the reality of ghosts, immaterial visitors from outer space, and shadowy detectives who could see into the minds of others through a kind of psychoanalytic x-ray vision. And when film adopted these audiophonic effects after 1927, it inherited much of the mystery and seeming omnipotence of the radio. By the 1930s, intimate, authoritative voices on radio and film in Europe and the US had become celebrities; they conducted church services, broadcast the news, and gave speeches that spread political propaganda.

The Italian Fascists and German National Socialists used radio and film extensively to spread their tribalistic, authoritarian propaganda. Radio was cheap, relatively easy to program, and reached most of their populations. Mussolini began broadcasting his open-air speeches in the mid-1920s and soon exercised complete control over the Italian airwaves. Recognizing the power of the medium, German propaganda minister Joseph Goebbels moved quickly to purge German national radio of all political opposition soon after the Nazis seized power. He approved funding for the subsidized sale or lease of cheap radio receivers for every German household and hired professionals to program a variety of entertainments as well as carrying Hitler's speeches, programs for German youth, and other vehicles for propaganda. Goebbels and the Nazis also invested in film propaganda. According to Leni Riefenstahl's *Memoir*, Hitler himself urged her to direct and produce *The Triumph of the Will*, the propaganda film that would forever define the seductive vision of Nazi tribalism and authoritarianism. Riefenstahl built on the success of Sergei Eisenstein's *Battleship Potemkin* (1926), the first film to merge the compelling artistry of montage with state propaganda and widely applauded in Europe as a modernist gem, despite its pro-communist, pro-egalitarian sympathies. She borrowed from Eisenstein for her images of euphoric national unity in *Triumph* and used his montage techniques to elevate the Führer as the necessary unifying symbol of German authoritarian power.

Triumph introduces Hitler as a god-like figure; at the opening of the film, he descends from the clouds in an airplane over Nuremberg and steps out of the plane at the airport to enthusiastic applause. As the movie proceeds, more time is devoted to Hitler's speeches and reaction shots from his devoted listeners. When the Führer urges purity, obedience, patriotism, and sacrifice to an audience of Hitler Youth, the young faces light up in response. A final rally at the climax of the film occurs at night in the huge outdoor amphitheater designed for the occasion by Albert Speer, with thousands of troops flanked by burning pyres and facing huge swastika banners behind the German eagle, with Hitler's podium in front of it. Riefenstahl's aerial cameras pan from this image of German solidarity and military might to fixate on the Führer, the focus of religious adoration. After numerous pledges of loyalty at the end of the rally, Rudolph Hess, caught in an apparent moment of relaxation, delivers his summation of the meaning of this ritual-like experience, "The party is Hitler and Hitler is Germany, just as Germany is Hitler" (Riefenstahl 1935). The subsequent history of Nazi Germany made it clear that many Germans were already beginning to agree. Riefenstahl's propaganda films, especially her religious depiction of tribalism and her elevation of Hilter for authoritarian worship in *The Truimph of the Will*, helped to make it so.

5.5 Postwar Attempts to Transcend Racist Tribalism

Although WW II locked in many of the same European antagonists against each other that had fought in the Great War, the racial and global dimensions of the second conflict distinguished the two and no doubt heightened the extent of the slaughter. By 1939, cultural-racist versions of nationalism had helped to elevate authoritarian fascists into power in Poland, Hungary, Italy, Spain, Greece, Rumania, and Yugoslavia, as well as Germany. On the world stage, Japan had also embraced racial nationalism and was using it to justify its military conquest of China in the 1930s. Although the Germans and the Japanese pursued racist war, the allies returned the lack of regard for the civilian lives of different "races" in the stereotypes of their war propaganda against "Nazis" and "Japs" and the US decision to place most west coast Japanese Americans into concentration camps. US racism played a role as well in the nuclear obliteration of the inhabitants of Hiroshima and Nagasaki in 1945. Overall, the death toll of WW II climbed to a staggering 70 million.

Revulsion against the racism of German and Japanese aggression, especially the Holocaust—shockingly apparent to all in the film footage taken by allied troops as they entered the camps to liberate their remaining occupants—helped to shape the peace. In his *The Invention of Humanity*, Stuurman reports that the war spurred the creation of the United Nations and, in one of its first major acts, the adoption of The Universal Declaration of Human Rights in 1948 by almost all of its member nations. Chaired by Eleanor Roosevelt at the height of her prestige as a progressive humanist, the coalition of nation-states and colonies authoring the Declaration celebrated universal rights for all human beings, modifying the masculine language of enlightenment rights but building on its underlying reasoning. The Declaration specifically bared discrimination on the basis of race and other tribal identifications and elevated the protection of these rights above the sovereignty of all nations. In theory, anyway, no polity could embrace and practice racism as a part of its politics, a major departure from the past hundred and fifty years of nationalism. Declaring that "everyone has the right to freedom of thought, conscience, and religion" (quoted by Stuurman 2017: 504), the 1948 Declaration completed the secularization of rights begun in the enlightenment. In addition to affirming equal rights for all women, the Universal Declaration also included language protecting girls from becoming child brides without their permission. The UN General Assembly overwhelmingly adopted the Declaration. Although no nation-state voted against it, there were three abstentions—from Saudi Arabia, the Union of South Africa, and the Soviet Union.

Situating the Universal Declaration "in the longer history of thinking about common humanity and equality since the seventeenth century" (Stuurman 2017: 507), Stuurman sees the 1948 statement as "a moral beacon buttressed by the collective authority of the world's sovereign states" (2017: 508). Endorsing a universalist notion of xenophilia, "it demonstrated that it was feasible to speak about the needs, expectations, and legitimate strivings of all the human beings on the planet without adopting a particular philosophical or religious stance" (508). People from widely different backgrounds had reached an accord on fundamental human rights, had condemned racism, and had even agreed on the importance of social, economic, and other egalitarian rights—albeit, without reaching consensus on how to enforce any of the declaration's standards. Although the advent of the cold war between the US and Soviet Union would stand in the way of implementing the goals of the Universal Declaration, the document remains an aspirational high point of xenophilia for its articulation of political and economic rights that could potentially unite most nations in an alliance against the depredations of tribalism, unjust hierarchies, authoritarianism, and—eventually—climate change.

Another impediment to realizing the aspirations of the Universal Declaration was the legacy of racist tribalism that haunted the process of decolonialization. Aware of the anti-colonial stance of most member states in the UN and also concerned about the high costs of sustaining imperialism, the European colonial powers began dismantling their empires. After dividing the Indian Raj between a predominately Muslim Pakistan and a mostly Hindu India, Great Britain allowed both polities to declare their independence in 1947. Burma and Ceylon also gained sovereignty in the same year. The Dutch left Indonesia in 1949, but the French stayed in Indo-China until their defeated army left Vietnam in 1954. The Japanese had replaced the British, French, and Americans as the major colonial power in China during the interwar years and their collapse in 1945 led to the full-scale resumption of the Chinese civil war. Soon after the Chinese Nationalist government retreated to Taiwan, the Communists under Mao Zedong declared victory, and the People's Republic of China was inaugurated in 1949. In the Middle East, after recognizing the independence of Syria, Lebanon, Iraq, and other Arab nations, the UK and France made good on their post-Holocaust promise to create Israel as a homeland for stateless European Jews. Most of the present nations in Africa gained sovereignty after 1960. A majority of the new nations sought to model their governance on democratic procedures, but for many egalitarian democracy would prove illusive.

As might have been expected, internal problems of racial and religious nationalism, kept in check by years of external imperial rule, erupted in many of the new nation-states after independence. The British division of India separated agricultural hinterlands from port cities and isolated millions of Hindus, Muslims, and Sikhs, who suddenly found themselves on the wrong side of hastily drawn borders in 1947. Nearly a million people died in the chaos that followed and India and Pakistan were soon at war over the disputed border of Kashmir in the North. In the Middle East, no sooner did the UN recognize Israel in 1948 than Arab countries on all sides of the nation invaded it. To block those states that sought to erase it from the map, Israel conquered and incorporated Arab lands on all sides of its initial borders. As a result, Israel is now unequally divided between predominately Jewish and Palestinian areas; thousands of Palestinians who live within Israeli borders have little democracy, few civil rights, and must depend upon Israeli and international aid for their survival. Ethnic and religious wars have plagued many African states as well, especially Nigeria, the most populous nation on the continent, which suffered four years of a bloody civil war, from 1967 to 1970. Since that time, numerous dictators from the three major Nigerian ethnic groups have ruled the nation, persecuting the many out-groups that struggle to survive in Nigeria. The Shell and Chevron companies that control the oil-rich Niger Delta region have worked hand-in-glove with Nigerian kleptocrats, reducing those minority ethnicities unfortunate enough to live in the Delta to near starvation. What all of these "postcolonial" polities in India, Israel, and Nigeria have in common is chronic and continuing tribalistic violence within their nation-states. Although racial nationalism has not caused a world war since 1939, it is clear that the implicit promise of an international system based on nation-states—a separate state for every ethnic nation—will continue to cause conflicts and killings.

Drawing hope from the founding of the UN and its articulation of universal rights, several western playwrights in the immediate postwar era wrote dramas celebrating the possibilities of humanity's ability to achieve equality and social justice. In France, Jean Giraudoux penned *The Madwoman of Chaillot*, a mostly euphoric allegory satirizing French capitalism and the quest for oil riches by advancing the whimsical idea that economic and political policy should follow the loving, egalitarian advice of four mad grandmothers to secure peace and equality among the rich and the poor. The corrupting power of wealth was a major theme of Swiss dramatist Friedrich Durrenmatt's *The Visit of the Old Lady*, as well. His grotesque, dysphoric play depicts the power of a rich woman to buy "justice" for herself after she transforms the poor, struggling citizens of a small town into her avenging murderers by bribing them with her immense wealth. Also reaching for universal significance was *The Crucible*, by Arthur Miller, which celebrates the ethical self-sacrifice of an everyman figure who breaks the power of a witch hunt in his small Puritan village in America to restore civil justice. The US postwar theater also included Brecht's *The Life of Galileo*, which sought to join egalitarian and scientific universals.

German writer-director Bertolt Brecht continues to be a major influence in contemporary European theater. Shaped by his experience as a medical orderly in the Great War and by the failure of revolutionary Communism to seize power in Germany in the early 1920s, Brecht gained commercial success with *The Threepenny Opera*, written with Kurt Weill in 1927, and then turned his back on bourgeois theater to embrace communist politics in 1928. In exile from Nazi Germany for most of the 1930s, Brecht sought refuge in Scandinavia and also made trips to the Soviet Union to seek theatrical and political allies. Stalin's purges and show trials in 1937–38, however, exiled or killed off several of his friends. Having planned to write about Galileo for several years, he completed a first draft quickly in November of 1938. In 1940, Brecht and his entourage took the trans-Siberian railroad across the Soviet Union and embarked for California, where he spent most of WW II, getting what film work he could in Hollywood. He stayed there long enough to help actor Charles Laughton mount a local production of *The Life of Galileo* in 1947. Although modestly successful, a remounted production in New York closed quickly at the end of the year. Called to testify before the House Un-American Activities Committee for his Communist loyalties, Brecht left the US for Europe and soon arrived in East Berlin, where he and his wife, Helene Weigel, convinced the new Communist authorities to subsidize a theater for the production of Marxist plays. Brecht and Weigel's theater, the Berliner Ensemble, produced his last version of *Galileo* in 1955, shortly before Brecht's death the following year.

I will use Marxist critic Darko Suvin's essay, "Heavenly food denied: *Life of Galileo*," as my jumping-off point for explaining the human and historical contradictions that Brecht animates in his drama. "I propose that the thematic domain of LG [i.e., *Life of Galileo*] is the pursuit of knowledge in class society," states Suvin. "Science should be knowledge making life for people easier, but it is alienated by ruthless pressures of authority into pure or specialized knowledge disregarding people's pursuit of happiness and the very existence of humanity" (Suvin 1994: 141). Although aware that he is living this central contradiction, Galileo fails to achieve his quest.

From Suvin's perspective, Brecht never finished revising *Galileo* and he traces key problems in the final version of the play to the dramatist's changing motives for writing and reworking it. The initial 1938–1939 version, written in Denmark, paints the great scientist primarily as a skillful tactician who pretends to be frightened of the Catholic Church's Inquisition so that he can live to write his famous *Discorsi* and send it via a trusted messenger to Holland for publication and distribution. In the second, American version, written in English for Laughton's performance in California in 1947, Brecht reversed his attitude toward his protagonist because of the horror of Hiroshima. The Bomb convinced him that scientists must work against the authority of the modern state for the betterment of humanity. From this point of view, Galileo's decision to recant his scientific principles and give in to the Inquisition was treason against humanity and his success in sneaking the *Discorsi* into publication more evidence of the collusion of Big Science with immoral authority. The final 1955 version of the play, written for performance by the Berliner Ensemble in East Berlin, contains most of the Laughton script and adds some minor changes to better realize Brecht's anti-Bomb position. Suvin points out, however, that it would have taken a much more radical revision of *Galileo* to fully reorient the play toward Brecht's understanding of the nuclear stand-off between the Soviet and Western blocs in the early Cold War. Nonetheless, notes Suvin, "The text strikes a deep chord in all spectators concerned with intellect versus power, with people caught between the institutionalizations of power and of knowledge" (Suvin 1994: 141). I agree. Big Science versus the nation-state continues as a relevant conflict today, especially with regard to the existential crisis of climate change. The climate crisis, however, also throws into sharp relief the shortcomings of Brecht's (and Suvin's) understanding of science and nature.

As Suvin points out, Scene 8, Galileo's debate with the Little Monk, "is compositionally and ethico-politically the central scene of the play" (Suvin 1994: 141). Before then, Galileo mounted a campaign to convince the

Church to discard Ptolemy's notion of an earth-centered solar system for the Copernican doctrine that the earth moves around the sun. In Scene 7, however, a powerful Cardinal tells him that the doctrine of Copernicus is "inane, absurd, and heretical" and admonishes Galileo to "relinquish this opinion" (Brecht 1972: 51). This massive defeat for the scientist sets up his ethical realizations in the Little Monk scene, allowing both to recognize the larger stakes of the conflict between Galileo and the Church. The Little Monk comes to Galileo a troubled man at the start of the scene because he is both a practicing Christian and a physicist who understands that Copernicus is correct. In a long, earnest speech, the Monk tells the scientist about his parents' difficult lives as poor peasants in the Campagna and says that he has decided to support the Church's decree against Copernicus because, otherwise, "what would be the good of the Holy Scripture which... demonstrates the necessity of all their sweat, patience, hunger, and submission, if it turns out to be full of errors" (1972: 56).

Galileo is sympathetic, but also excited. For the first time in the play, he recognizes the connection between the taxes paid by the peasants of the Campagna and the Church's need to maintain its supremacy on earth as well as in the heavens, specifically in the Thirty Year's War it is waging in German lands against Protestantism. After dismissing the limiting definition of virtue behind the Monk's pleading, Galileo challenges him: "You describe your peasants in the Campagna as if they were moss on their huts.... Damn it, I see the divine patience of your people, but where is their divine wrath?" (Brecht 1972: 58). Then he decides to use the Monk's curiosity to spark a more radical response. He throws some scientific manuscripts in front of the Monk as a cook would throw meat scraps to a dog and declares "there you'll find the reasons for the ocean's tides" (1972: 59), daring the Monk to read them. Unable to avoid the temptation, the Monk snatches them up and starts devouring their contents. "An apple from the tree of knowledge," replies Galileo gleefully. "He gobbles it up. He'll be damned for all eternity, but he's got to bolt it down, the hapless glutton" (59). As Suvin understands, this climax to the scene relies on one of Brecht's master metaphors for his drama, which likens the quest for scientific knowledge to our natural appetite for food. Both are universal to the human condition, believes Brecht, and both may result in progress or destruction for humankind.

The implicit promise of Scene 8—translating Galileo's teachings into revolutionary, egalitarian power for the people—is suggested most forcefully two scenes later, in the carnival scene. As occurs at the top of several scenes in *Galileo*, Scene 10 begins with a placard lowered for the audience, which reads: "In the course of the next ten years Galileo's doctrine is disseminated among the common people. Pamphleteers and ballad singers everywhere seize upon the new ideas. In the carnival of 1632, the guilds in many Italian cities take astronomy as the theme for their carnival processions" (Brecht 1972: 70). The procession in Scene 10 features a crowd and a ballad singer who celebrates the topsy-turvy societal implications of Galileo's teachings. One verse asks, "Good people, what shall come to pass / If Galileo's teachings spread? / The server will not serve at mass / No servant girl will make the bed." More alarming, in the next verse, "The carpenters take wood and build / Houses for themselves, not pews / And members of the cobblers' guild / Now walk around in shoes!" (1972: 72). Scene 10 ends with the entrance of a life-sized, stuffed figure of Galileo, "which bows to the audience. In front of it a child displays a giant open Bible with crossed out pages" (73). The Ballad Singer introduces the figure as Galileo, "the Bible smasher!" and the crowd laughs and applauds (73).

Unfortunately, argues Suvin, Brecht did little more in his revision of the play to indicate the potential revolutionary impact of Galileo's ideas. In Scene 14, the climax of the play, the dramatist has the scientist blame himself for giving in to the Inquisition and turning his back on the people when he might have made a difference in their lives. Suvin asks, however, "just who and where [in the play] are the rising lower classes?" (Suvin 1994: 149). It's an apt question. The short Carnival scene comes and goes without establishing a relationship between the revelers and the actual Galileo and no other scene in the play even depicts possible lower class allies for him. Unlike the upper class, fleshed out with several churchmen, political rulers, and their emissaries, the lower classes primarily exist in the abstract. As Suvin concludes, "The stupendous dialogue with the Little Monk about the Campagna peasants does not quite compensate for this lack of [embodied] connection" (1994: 149) in *Galileo*.

While I agree with Suvin on this point, it was probably less important than he implies for the relative success of Brecht's productions of his play. The dramatist did most of his work on the Laughton version of *Galileo* in the second half of 1945, which places that revision directly in the shadow of Hiroshima, obliterated on August 6 of that year. From then until July of 1947, when the play was finally produced by Laughton, there were extensive discussions in the US press about what to do with atomic bombs. Many Americans, including J. Robert Oppenheimer who had directed the Manhattan Project, feared that other nations would also build and deploy atomic weapons and that this would lead to an arms race that would soon destroy life on Earth. President Truman formed a commission to investigate possibilities for the international control of nuclear energy and the Acheson-Lilienthal Report, authored partly by Oppenheimer, recommended that the new UN establish a regimen of inspections that would police all nations to ensure that no new atomic bombs were manufactured, after which the US would turn over all of its bombs to a UN atomic energy commission. The US presented this plan at the first meeting of the UN in 1946, but the Soviet Union, fearing that the US could not be trusted and resenting inspections as a potential violation of its sovereignty, refused to bring it to a vote in the Security Council. By the middle of 1947, the UN plan was dead and a new arms race, fueled by the start of the Cold War, had begun. After 1949, when the Soviets got The Bomb, the two armed camps would not agree to limit the number of nuclear warheads until 1972, when both sides signed the first START treaty.

The 1945–1947 international context is important because Brecht knew that US audiences watching his drama would understand that the "new age" announced by Galileo at the start of the play signaled the start of the atomic age and they would be alert to the drama's implications regarding the future for life on an Earth. Indeed, there is substantial evidence that Brecht revised the play with this audience expectation in mind. It would also have been obvious to Brecht during this time that a communist revolution led by an alliance of workers and scientists was not going to occur in the postwar US. Not only had Roosevelt's New Deal turned most workers toward an accommodation with capitalism, but nuclear scientists like Oppenheimer, along with other US liberals and socialists inclined toward seeking international peace, faced significant opposition from those in the military and among the leaders of major corporations—a powerful alliance for an arms race that would later be named "the military- industrial complex" by President Eisenhower. There was also the intransigence of the Soviet Union at the UN for Brecht to consider. Already aware of Stalin's authoritarian amorality, he had no reason to believe that Stalin's brand of Marxism would ever favor a genuine revolution from below. Given this situation, it is not difficult to believe that Brecht decided only to acknowledge the revolutionary potential of "the people" in the abstract, but to aim the final scenes of his revised *Galileo* against those military-industrial elites and scientists who were opposing the present-day Oppenheimer-Galileos pushing for peace. Although a Marxist proletariat was unlikely to rise against the US or the Soviet Union in the late 1940s, Brecht understood that his play might animate ethically minded scientists and others to stop the spread of nuclear weapons through the UN.

Consequently, from Brecht's perspective, Galileo's failure to forthrightly oppose the warmongers of his own time—unlike the principled stand of Oppenheimer and his scientific allies—deserved condemnation. Scenes 1 through 13 of the play depict Galileo's two failed attempts over roughly thirty years to win papal approval for his new scientific discoveries in order to better the lives of the common folks of Europe. By the climactic scene of the play, Brecht's Galileo is disgusted with himself for having given in to these pressures.

At the beginning of Scene 14, a peasant carries two plucked geese into Galileo's apartments and the scientist, now old and nearly blind, tells his daughter to prepare one of them for dinner. It soon appears, though, that Galileo's appetite for new knowledge is not as keen as his appetite for food. An early ally and now an enemy, Andrea Sarti, enters to tell Galileo he has come to visit him at the request of a scientist in Amsterdam who has asked about his health. Sarti remains cool to the scientist, however, until Galileo tells his former student that he has now finished his *Discorsi*, a new scientific treatise on the physics of motion, and he wants Andrea to take it with him to the Netherlands for publication. Overjoyed, Andrea apologizes for ever doubting Galileo's loyalty to his scientific calling. Even after Galileo tells him that he recanted to the Inquisition to save himself from physical pain, Sarti continues to justify his actions, stating, "In science only one thing counts: contribution to knowledge." "Welcome to the gutter, brother in science and cousin in treason," Galileo retorts. A few lines later, he adds that although he no longer considers himself to be a scientist, "I believe I can give you a few hints about the science you are devoting yourself to" (Brecht 1972: 93).

With that, Galileo, "in lecture style, hands folded over his paunch" (1972: 93), delivers the moral of Brecht's drama. Science demands courage, he says, which leads to the possibility of what could be termed euphoric victories but more often to dysphoric failures. Because "princes, landlords, and priests keep the majority of the people in a pearly haze of superstition and outworn words to cover up their own machinations," the people need our truths (93). "But can we turn our backs on the people and still remain scientists?" asks Galileo, rhetorically. "What are you scientists working for? To my mind, the only purpose of science is to lighten the toil of human existence. If scientists... confine themselves to the accumulation of knowledge for the sake of knowledge, science will be crippled and your new machines will only mean new hardships" (94). Then, in a comment pointing directly at Hiroshima, Galileo states, "The gulf between you and humanity may one day be so wide that the response to your exultation about some new achievement will be a universal outcry of horror." Next, Galileo turns reflective about recanting: "If I had held out, scientists might have developed something like the physicians' Hippocratic oath, the vow to use their knowledge only for the good of mankind. As things stand now, the best we can hope for is a generation of inventive dwarfs who can be hired for any purpose." Guilt-ridden, he concludes, "I have betrayed my calling. A man who does what I have done, cannot be tolerated in the ranks of science" (94).

In terms of its dominant emotions and generic appeal, Brecht's play ends as a Jeremiad, a denunciation of sin and corruption in the Temple of Science. Galileo, driven more by his appetites than by his principles, certainly makes for a divided, even Falstaffian Jeremiah, but that is a part of his (and the play's) charm and persuasiveness. Successful Jeremiads seek to inspire guilt and shame in their listeners—the recognition, in this case, that the challenges posed by modern polities demand much more of us than most scientists have given so far. Specifically, Brecht's Galileo, despite his failings, taps into audience guilt by underlining our compromises with political power, compromises that could lead to another nuclear war and the end of humankind. Guilt is a dysphoric emotion, akin to disgust; it is meant to stir us out of self-satisfaction and into altruistic or at least cooperative action. Of course Galileo also rucks up guilt and disgust in himself, lending the play more of a pessimistic than a comic ending. On balance, then, the play is more dysphoric than euphoric.

Although Brecht's point of view about science and political power still has much appeal, it continues the mind–emotion split that helped to legitimate and sustain the rise of capitalism. On the face of it, Brecht's call for a Hippocratic oath among scientists to ensure that they serve the good of humankind seems straightforward and necessary. As in all social institutions, scientific practice demands ethical responsibility and this, in turn, requires occasional Jeremiads to enforce appropriate institutional norms. According to Suvin, each of Brecht's versions of Galileo "strikes a deep chord in all spectators concerned with intellect versus power, with people caught between the institutionalizations of power and of knowledge" (Suvin: 141). Suvin is correct that Brecht places science on the side of "intellect," but this suggests that "power" is aligned with its usual opposite, emotion.

The cognitive realities of individual "intellect," however, especially the importance of empathy and the emotions in arriving at scientific knowledge, undercut the relevance of Brecht's play for current coevolutionary understandings of scientific practice. For most of his career, Brecht divided rational perception from emotional response and nearly always preferred the former to the latter. The most infamous instance of this duality was his rejection of what he understood to be empathy as the basis for spectator response in the theater. He inherited his ideas about *Einfühlung* (the German word now translated as empathy) from nineteenth-century German Romantics, who used the word and its cognates to mean the ability of individuals to project themselves into the spirit of another person or natural object. From the Romantics' perspective, this notion of identification and mystical merging involved conscious projection and the temporary loss of the self in another person or object. Following the Great War, when Brecht began writing and directing, German Expressionism was at the height of its popularity in Berlin. From Brecht's point of view, however, Expressionist actors too easily lost themselves in their characters and their audiences enjoyed immersing themselves in emotional stories that allowed them to forget about social and economic realities; for Brecht after his conversion to Marxism, Expressionism was *Einfühlung* run amok. He rejected what he took to be the goal of Romantic identification and mystification for the precepts of Marxist science. Following the enlightenment principles of Descartes, this understanding of science separated the mind from emotions and humanity from nature and it encouraged its advocates to use science for the advancement of socialist progress and a communist revolution.

From the late 1920s until the mid-'30s, Brecht sought to build his ideal theater on the basis of rational, rather than empathic response. Fearing that *Einfühlung* would cause spectators to lose themselves emotionally, he argued that empathy "wears down the capacity for action" in the audience (Brecht quoted in Willett, 1964: 37). Under the spell of empathy, wrote Brecht, "nobody will learn any lessons" (1964: 26) about politics and economics. These prejudices were scientifically baseless, but they were also widely shared. When he returned to playwriting after the mid-1930s while living in exile, however, Brecht found that his interest in writing about complex characters forced to make difficult moral decisions often involved him in portraying these figures in ways that would tempt spectators to identify emotionally with them. In Scene 14 of *Galileo*, for instance, Brecht provides two paths for audiences to understand his disgust with science—the path of rational argumentation via Galileo's lecture and a path that simply invited the audience to empathize with Galileo's self-disgust. Audiences could take both paths, of course, and arrive at the same conclusion, but Brecht remained reluctant to recognize that empathy, by itself, could be a path to self-knowledge.

More problematic for the narrative structure of *Galileo* was Brecht's commitment to Cartesian rationalism as the primary mode of scientific discovery. Although the play celebrates Galileo's and the Little Monk's curiosity, Brecht's commitment to enlightenment reason hobbled *Galileo's* understanding of the social basis of scientific discovery. Contemporary scientists and philosophers have been poking at the problem of reason for several decades now, mostly with inconclusive results. A recent book by Hugo Mercier and Dan Sperber, *The Enigma of Reason* (2017), however, brings an insightful coevolutionary and interactionist perspective to bear on the problematics of rationality and empathy. As many historians have pointed out, there are copious examples of famous scientists reasoning badly; Newton embraced alchemy, Pauling obsessed on the virtues of vitamin C, and Jefferson believed he had discovered the scientific basis of African American inferiority. Mercier and Sperber ask how these apparently brilliant men could have made such stupid mistakes. They begin by turning to the likely evolutionary origins of reasoning and place this mode of cognition in the context of our species' ultra-sociality. From this perspective, reasoning did not arise as a means for individual members of our species to make up their minds in isolation. Indeed, anyone can pile up "reasons" to justify his or her conclusions, a process now understood by most scientists as "confirmation bias." Rather, Mercier and Sperber demonstrate that reasoning works most effectively in the context of interactive conversation and debate; we reason best by sifting good ideas from bad ones in the process of assessing other's conclusions, not our own. After the invention of language, this process would have had obvious survival value for our hunter-gatherer ancestors. Those who could empathize with others' points of view and persuasively marshal their arguments to support the good of the band would have increased their prestige among band members. As Mercier and Sperber point out in a section entitled, "Science Makes the Best Use of Argumentation," conversations in lab meetings and through later publications are often a fruitful means of extending and elaborating such discussions.

Brecht's *Galileo*, however, does not dramatize this side of scientific exploration. Although far from the "Genius Scientist" biopics that Hollywood was churning out at the time, the play does not show Galileo in the midst of scientific interactions with his colleagues. The audience occasionally learns about his wide reading of others' insights, but nearly all of the scenes in which science is under discussion depict Galileo either at the beginning of one phase of his discoveries or at the end of them, when he is explaining what he has learned to Andrea or other characters. Notably, scene 14, the penultimate episode of *Galileo*, places Brecht's disgusted genius in "lecture" mode, the opposite of interactive give-and-take. In short, Brecht depicts Galileo primarily as an isolated genius in the Laughton version, all the more to be pitied and/or celebrated because of his isolation. Although this strategy is dramatically appealing (and may even be historically accurate), it misleads audiences about the process of scientific discovery and validation and takes away from Galileo as a potentially interactive and empathic model for contemporary scientists.

Evidently, the version of Marxism followed by Brecht owed too much to the mind–body split in Descartes for Brecht to easily embrace a mode of scientific reasoning and explanation that would now be recognized as compatible with the aspirational understandings of ecology and evolution, which some progressives are struggling to adopt today. In his introduction to *Capitalism in the Web of Life*, Jason Moore recognizes that Cartesian dualities—especially mind/body and its correlate, human society/primitive nature—continue to pervade our thinking: "Our scholarly vocabularies, even after four decades of Green Thought, are still contained within – and constrained by – an essentially Cartesian notion of nature-society interaction" (Moore 2015: 22). Despite the need for a new vocabulary to approach the interpenetrations of "body" and "mind," plus "nature" and "society," we can begin to see what these dynamics might look like in a few recent plays that focus on scientific investigation and discovery in the midst of historical oppressions. In this regard, US playwright Lauren Gunderson's female scientists often enjoy sparring with their fellow researchers to puzzle out scientific insights. See, for example, *Emilie: La Marquise Du Châtelet Defends Her Life Tonight* (2010), which pits the Marquise against Voltaire and the legacy of enlightenment science. Unfortunately, Brechtian Marxism owed too much to

Descartes to embrace a more dynamic and democratic mode of scientific reasoning and explanation. In current campaigns to make science more responsible for Mother Nature's and humanity's needs, our Pleistocene legacy of empathy and egalitarianism may be a better guide to help us toward an understanding of scientific practice that supports group discovery, equality, cooperation, and xenophilia.

REFERENCES

Anderson, Benedict. 1991. *Imagined Communities: Reflections on the Origins and Spread of Nationalism*, 2nd ed. London: Verso.

Brecht, Bertolt. 1972. Life of Galileo. In *Bertolt Brecht Collected Plays*, vol. 5, ed. Manheim, Ralph and Willett, John. New York: Vintage Books.

Cohen, Mitchell. 2017. *The Politics of Opera: A History from Monteverdi to Mozart*. Princeton, NJ: Princeton University Press.

Crosby, Alfred. 1972. *The Columbian Exchange: Biological and Cultural Consequences of 1492*.

Gerwarth, Robert. 2016. *The Vanquished: Why the First World War Failed to End*. New York: Farrar, Straus & Giroux.

Hogan, Patrick Colm. 2018. *Literature and Emotion*. New York: Routledge.

Judson, Pieter M. 2016. *The Habsburg Empire: A New History*. Cambridge, MA: Belknap Press.

Lieberman, Debra. 2018. *Objection: Disgust, Morality, and the Law*. Oxford: Oxford University Press.

Mercier, Hugo, and Dan Sperber. 2017. *The Enigma of Reason*. Cambridge, MA: Harvard Univ. Press.

Poe, Marshall T. 2011. *A History of Communications: Media and Society from the Evolution of Speech to the Internet*. Cambridge: Cambridge Univ Press.

Mitchell, Timothy. 2011. *Carbon Democracy: Political Power in the Age of Oil*. New York: Verso.

Moore, Jason W. 2015. *Capitalism in the Web of Life: Ecology and Accumulation of Capital*. New York: Verso.

Riefenstahl, Leni. 1935. *The Triumph of the Will*.

Stuurman, Siep. 2017. *The Invention of Humanity: Equality and Cultural Difference in World History*. Cambridge, MA: Harvard University Press.

Suvin, Darko. 1994. Heavenly Food Denied: *Life of Galileo*. In *The Cambridge Companion to Brecht*, ed. P. Thomson and G. Sacks. Cambridge: Cambridge University Press.

Wilkerson, Isabel. 2020. *Caste: The Origins of Our Discontents*. New York: Random House.

Willett, John. 1964. *Brecht on Theatre*. London: Metheun.

PART III

Neoliberal Capitalism and Political Disintegration in the US

Having established major commonalities in politics and drama over the *longue durée* of our species' history, I turn in Part III to the role played by dramatic performances in helping to change a single polity over the course of seventy-five years. Deploying retrodiction and other cliodynamic strategies, Turchin's *Ages of Discord: A Structural-Demographic Analysis of American History* reveals that contemporary US society and its polity have been fragmenting since the mid-1960s. His book begins with a flashback to the surrender of Ft. Sumter in 1861, which signalled the catastrophic failure of the US constitutional system and triggered the Civil War. As Turchin puts it, "What is particularly astounding is how myopic the American political leaders and their supporters were on the eve of the Civil War, especially those from the Southern states" (Turchin 2016: 3). He warns that the Civil War should have taught us "that large-scale complex societies are actually fragile, and that a descent into a civil war can be rapid"—a lesson that, "150 years later ... has been thoroughly forgotten" (2016: 4). Based on his research, Turchin predicted in a 2010 article in *Nature* that the US would suffer major social unrest around 2020. Publishing *Ages of Discord* in 2016, Turchin states that we now know enough about historical dynamics—even though that knowledge "is by no means complete"—"to be worried, very worried about the direction in which the United States is moving" (4).

The US is now nearing the end of what Turchin calls its second complete cycle of sociopolitical integration and disintegration. His structural-demographic theory (SDT) identifies long-term cycles of political stability and instability, usually lasting from a hundred and twenty to two hundred years, in such disparate, agriculturally based polities as the Roman Republic, Bourbon France in the years before the Revolution, and China over the two-thousand years of its empire. Each cycle consists of integrative and disintegrative phases, roughly equal in length, when the main indicators of political instability and social discord are either decreasing (the integrative phase) or increasing (disintegrative). Turchin found that his SDT—based primarily on synthesized ideas from Marx, Weber, Malthus, and others and modified by masses of historical data—worked well to retrodict changes in the early years of US history. He acknowledges in *Ages of Discord* that he modified the theory for the period between 1840 and 2010, due to changes forced by industrialization and, later, by the dynamics of neoliberalism and post-industrialization.

Turchin's indicators demonstrate that the first full cycle of integration and disintegration in the US ended around 1900. After that, a second integrative phase began to track upward with progressive reforms, the national unification required for US entry into the Great War, New Deal economic changes, mobilization for WW II, and the postwar economic boom. The cycle began to shift again in the mid-1960s with the fragmentation of the Civil Rights struggle, urban riots in the North, protests against the Vietnam War, and capitalism's attacks on the power of labor unions. By 1980, his SDT variables indicated that the US had entered a disintegrative cycle. "I think it appropriate to refer to this turning point as the Reagan Era Trend Reversal," says Turchin, "because the presidencies [of Reagan and GHW Bush] … were the period when the new economic and social regime became obvious to all observers" (2016: 208). He is quick to add that he is not blaming Reagan and Bush for these declines in indicators of social cohesion; no one person or one group can be held responsible for them. Rather, Reagan's presidency was largely a consequence, and became a symbol, of these deeper structural shifts.

Turchin used several material proxies to tabulate his structural variables for this reversal. (See *Ages of Discord*, Table 11.1, page 200, for a listing of all 29 proxies that Turchin deployed to get to 15 structural variables.) For the "well-being" variable, for example, Turchin drew on historical indices of relative wages, physical stature, average age at marriage, and life expectancy. Note that three of these four indices are straightforward biological measurements of well-being. For the variable of "labor oversupply," Turchin turned to three indicators with reliable historical data: proportion of foreign-born, foreign trade balance, and the labor demand/supply ratio. Other structural variables for the reversal include economic well-being, social well-being, economic inequality, social cooperation, and trust in government.

Examining three general areas of social demographics, Turchin's *Ages of Discord* charts the structural dynamics driving this post-1965 shift in terms of the general population, elite groups, and the government. For the general population, he finds labor oversupply increasing, relative wages (as a percent of GDP) declining, and economic inequality increasing in the 1965–1980 period. Regarding elite groups, Turchin discovered that the elite grew as a percentage of the population, that its growth spawned intra-elite competition, and that its willingness to participate in social cooperation declined between 1965 and 1980. He tracked the number of lawyers as a percent of the general population, for instance, and found that their numbers began to increase rapidly in the late 1970s, along with a rise in political polarization among elites. In general, as the wealth and power of elite groups rose, social well-being and economic and health indicators for the general population fell. Added to these three demographic groups was a general index of the political forces of instability, which include radical ideologies, terrorism, and riots, plus revolution and civil war. Together, these indicated that sociopolitical stability, political legitimacy, and patriotism reversed from the mid-1960s through the 1970s. While each of these structural variables is significant, Turchin emphasizes that it is the dynamic interaction of all of them that creates the cycles.

While *Ages of Discord* has discovered significant indications of social and political dissolution in the US since 1965, it is clear that Turchin's demographic analysis does not map easily onto the indicators that I have relied upon thus far in *Drama, Politics, and Evolution* to examine our coevolutionary inheritance of political practices and tensions from the Pleistocene. Turchin's SDT incorporates some indictors for changing norms of cooperation and it is certainly interested in tracking matters of perceived equality and tribalism. But his indicators do not understand matters of legitimate authority as a tension between the social play of domination and prestige. Nor is xenophilia necessarily the opposite of tribal loyalties in his method, as I have defined and discussed that tension. In short, although both of us are interested in the general problem of social cohesion, I deployed categories from my sources that often overlap with but are not coterminous with Turchin's indicators and methods. This does not mean that our general conclusions about the decline of social cohesion in the US over the last fifty years are necessarily in conflict, of course; for the most part, as will be evident, the facts and reasoning I used to arrive at my explanation simply emphasize different aspects of similar factors and take a different methodological path to arrive at conclusions.

Accordingly, I will stay with the terms of analysis I introduced in Chapter 1—social norms, cooperation, altruism, and the tensions between domination and prestige, hierarchy and equality, and tribalism and xenophilia—rather than switch to Turchin's categories and methods in the three chapters that follow. I will use my previous terminology to frame the work of several historians, political scientists, investigative reporters, and other reliable sources who have taken several different approaches to the political and performative history of the 1945–2020 period in the US. And of course I will continue to deploy Whitehouse's methodology to explain why certain kinds of dysphoric and euphoric dramas, nearly all of them on film, gained popularity in the US between the 1950s and 2019. These include: *Dr. Strangelove* (1964), *Night of the Living Dead* (1968), *The Godfather* (1972), *Do the Right Thing* (1989), *Independence Day* (1996), *Avatar* (2009), *The Dark Knight Trilogy* (2005–'12), *Lincoln* (2012), *Django Unchained* (2012), *The Hunger Games* series (2012–'15), and *Joker* (2019).

Focused on sociopolitical cohesion, I will look primarily at domestic politics, not foreign affairs, unless foreign wars and difficulties impacted domestic concerns. As noted in Chapter 1, I will rely primarily upon material indicators and social scientific data rather than ideological analysis, though ideology will play a role. Occasionally, my perspectives will overlap with Turchin's, but none of my other sources consistently turn to big data and cliodynamic analysis. My goal is to discover whether other investigators using different methods were reaching more or less the same conclusions about the downward spiral of social cohesion in recent US history. Unfortunately for the US, a great many diverse sources have confirmed Turchin's pessimistic conclusions.

CHAPTER 6

Postwar Hegemony and the Reagan Reversal, 1945–1985

This chapter primarily explores the quick rise of American world hegemony after WWII, the effects of American militarism on the nation, and dramatic trend reversals in social cohesion between 1965 and 1980, evident as well in the popularity of dysphoric film genres at the time. Unlike most imperialists, the American elite was reluctant to acknowledge publicly that they controlled much of the world for their own economic benefit during the postwar era. It was easier to explain US power to the American people as a defense against world communism, which, along with the inherent superiority of capitalism and the will of God, emerged as the primary justifications for the American empire. In truth, however, the US planned for world hegemony during WWII and used the early years of the cold war with the Soviet Union to solidify its gains. American world domination would not have been possible, however, had not Europe—principally Germany, Great Britain, and France—and its colonies been decimated by two world wars between 1914 and 1945. Escaping the worst economic effects of war, of course, was largely a matter of fortunate geography for the US; no European or Asian army in either war could easily invade or even bomb the US mainland. Unsurprisingly, American politicians in the postwar era never credited geographic luck as a primary reason for the nation's economic prosperity.

© The Author(s), under exclusive license to Springer Nature
Switzerland AG 2021
B. McConachie, *Drama, Politics, and Evolution*,
Cognitive Studies in Literature and Performance,
https://doi.org/10.1007/978-3-030-81377-2_6

6.1 The Domestic Consequences of World Hegemony, 1945–1975

In 1944, the US laid the foundation for its postwar economic dominance at a meeting with representatives of the other Allied governments at Bretton Woods, a resort in New Hampshire. By this point in the war, American loans were bankrolling the war effort against Germany and Japan and US allies had little leverage to object to the financial structures proposed by the US. Under the Bretton Woods Agreement, all postwar currencies were to be pegged to the US dollar, the International Monetary Fund and the World Bank (both to be dominated by US appointees) would regulate world financial affairs, and the General Agreement of Tariffs and Trade (GATT) would open all markets to free trade. With many European countries and their colonies in ruins after the war, GATT allowed US capitalists to import cheap raw materials from around the world and to increase their sales of exported goods abroad with little competition from foreign capitalists. Unlike previous world hegemons, the GATT and its results allowed the US to dominate world trade without the need to hold and maintain colonies around the globe. In the language of Jason Moore, the US controlled three of the "Four Cheaps" in the postwar world economy—cheap food, cheap energy, and cheap raw materials. Bretton Woods functioned smoothly until the late 1960s, ensuring free trade, the growth of American manufacturing, and high American profits.

American patriots had promoted tribalistic nationalism during WWII and this was easily revived less than two years after the end of the fighting, when it seemed to many that Soviet Communists were taking over Eastern and Southern European countries to expand their power. In April of 1947, President Truman mobilized support for the Marshall Plan, a long-term program involving loans, direct assistance, and mutual commitments designed to end the economic devastation of Europe. In addition to challenging the Communists in Europe, Truman recognized that European recovery would help to promote and extend the postwar prosperity of the US. By requiring that countries accepting aid also purchase American petroleum, Truman used the Marshall Plan to move the Western European economy away from its reliance on coal and toward oil. Later in 1947, Truman, along with security-minded Democrats and Republicans, also passed the National Security Act. The act created three agencies that would play a key role in the emerging cold war: The Central

Intelligence Agency (CIA) to coordinate spying outside of the US, the National Security Resources Board to link the needs of the armed services with universities and corporations, and the National Security Council (NSC) to advise the President on all matters relating to the security of the nation. Through these agencies, the law empowered the President to mount external propaganda campaigns, destabilize enemy political regimes, and conduct covert warfare to protect American interests. As political philosopher Frederick M. Dolan noted, the vast expansion of presidential authority implicit in the 1947 Act signaled a change in "the constitutional regime" of the US (Dolan 1994: 60). After 1947, the American people and their representatives had no legal access to knowledge deemed "top secret" by the security establishment. The law would have important consequences for the imperial reach of American power and opposition to it, both in the US and abroad.

From Turchin's perspective, the 1947 law may also be understood as the politically "realist" response of a rising empire to the threat of war. Like previous military powers in world history, the US added military and economic complexity to its polity to enable the nation to prepare for what many analysts understood as a likely military contest with the Soviet Union; preparations for war drove higher levels of social and economic complexity. Behind the 1947 act was recognition of the need for better spying, better education, and better integration of the nation's economy with its war-making capabilities. This last level of complexity would lead President Eisenhower to warn in his Farewell Address in 1961 against the rise of a "military-industrial complex" in the US that threatened to gain political power, corrupt elections and governance, and dominate democratic decision-making at the national level. As several historians have related, Eisenhower's warning was already too late. Most Americans willingly embraced the marriage of corporate profits and military might. And they fought to build and maintain military bases, armament industrial sites, and educational facilities that directly supported US militarism in their congressional districts. After the Berlin Airlift and the formation of the North Atlantic Treaty Organization (NATO) in 1949 to stop Soviet aggression in Europe, these investments seemed worth it.

From 1947 into the mid-1970s, the US pursued a policy of containing "Communism," understood by most Americans and their leaders as a monolithic international movement. By 1947, the House Un-American Activities Committee (HUAC) was investigating Communist influence in the film industry, which resulted in Hollywood's decision to blacklist suspected directors, writers, actors, and others. The "one world" cosmopolitan vision of J. Robert Oppenheimer and other like-minded atomic scientists was clearly losing popular support. When the Chinese Communists pushed the Nationalists out of mainland China in 1949 and declared victory in their long civil war, many Americans naively saw the Communist victory through this nationalistic lens. With the world apparently divided between "freedom-lovers" and "communists," the accusation, "Who lost China?" became a rallying cry for conspiracy-minded Republicans eager to pin the blame on Truman and the Democrats for the sudden spread of red ink adjacent to the Soviet Union on world maps. As the hunt for Communists in the State Department grew, the news in September of 1949 that the Soviets had exploded an atomic bomb shook Americans out of their expectation of continued military invincibility. More suspicions about Russian spies and US subversives rattled the American public when the Communist North Koreans sent troops into South Korea in June of 1950 and Truman ordered America to aid the South Koreans as a part of a UN "police action."

The conservative tribalism embraced by republican anti-Communists was eagerly defended by the Reverend Billy Graham and others on the Protestant Religious Right, who never tired of denouncing "godless Communism." Not surprisingly, given Tuschman's conclusions about tribalism, ethnocentric beliefs and religious faith were also aligned with conservative attitudes toward sexuality. Most fervent anti-Communists embraced a rigid heteronormativity that demanded the persecution and firing of suspected homosexuals in the State Department and elsewhere in the government. For many national leaders, however, anti-Communism had become a two-edged sword. While it justified the expansion of American power and influence around the world, it also limited the flexibility of political maneuvering in domestic as well as foreign affairs. Unfortunately, the primary political lesson learned from the anti-Communists was never to expose yourself to the accusation that you were "soft on Communism." This "lesson" would haunt US policy during the Cuban Missile Crisis and throughout the buildup and war in Vietnam, from 1963 through 1974. By 1954, the US was already paying three-quarters of the French war costs in what was still French Indochina.

The Cuban Missile Crisis of 1962 grew out of the national policy to contain the spread of Communism. President Kennedy learned in October that the Soviet Union had started to construct missile bases in Cuba, its island ally, imposed a naval blockade to stop the buildup, and warned the Soviets that the continued presence of missiles in Cuba would lead to nuclear war. Following the official US policy of MAD—that is "mutually assured destruction"—the President ordered twenty-three nuclear-armed B-52s to orbit points within striking distance of the Soviet Union. After a couple of days of white-knuckled waiting, however, Kennedy accepted Premier Khrushchev's offer to remove the missiles if the US promised never to invade Cuba. As Kennedy hoped, most analysts and reporters in the West interpreted the resolution of the crisis as a victory for the US. Instead of blaming Kennedy for taking the world to the brink of nuclear destruction, most Americans saw the events through the moralistic lens of a Hollywood western. As in *High Noon* (1952) with Gary Cooper, a brave US Marshall had protected America from the bad guys with his superior nerve, courage, and firepower. In the showdown with evil, Khrushchev had "blinked first," a widely reported remark attributed to Secretary of State Dean Rusk (Kennedy 1971: 77).

This perception of American victory in the missile crisis was ironically echoed a year and a half later in film director Stanley Kubrick's satirical treatment of a cold war nuclear apocalypse, *Dr. Strangelove or: How I Learned to Stop Worrying and Love the Bomb*. Released in 1964, the film reminded many critics and viewers of the catastrophe humankind had barely escaped. Near the climax of the movie, soon before an American bomber drops an H-bomb on the Soviet Union, T.J. "King" Kong, the commander of the airplane (played with a rich Texas accent by Slim Pickins), announces: "Well boys, this is it: Nuclear combat, toe-to-toe with the russkies" (Kubrick 1964).

Set in the nearly all-male and White world of US generals, diplomats, scientists, and politicians in a Pentagon War Room, on the Airforce base of a general who goes insane and decides to start a nuclear war with the Soviet Union, and inside a B-52 bomber that successfully outmaneuvers Russian surface-to-air missiles to drop its nuclear payload, *Dr. Strangelove* centers on male sexual confusions and fantasies. Tellingly, the only female in the film is General Buck Turgidson's girlfriend, a magazine fold-out model, who calls him on the phone—interrupting his meeting with the Soviet Ambassador and others to save the world from looming nuclear disaster—in order to lure the turgid general back to bed. Most of the other alpha male types have similarly suggestive names: Jack D. Ripper (the insane USAF general), Merkin Muffley (a reference that suggests the impotence of the US President), Alexei de Sadeski (Soviet Ambassador), the aforementioned "King" Kong, and Dr. Strangelove himself (a deformed former Nazi scientist in love with power). In the course of the narrative, spectators learn that the Russians have deployed a "Doomsday machine" to ensure MAD if attacked by a single nuclear weapon and the film ends with footage from the nuclear tests of multiple exploding H-bombs.

While awaiting Doomsday, the men in the War Room, led by Dr. Strangelove, fantasize about the fertile young women they would like to take with them into bunker mineshafts in order to repopulate the world during the nuclear winter to follow. Major Kong metaphorically fulfills their libidinous daydreams by mounting an enormous sperm-shaped H-bomb in the bomb-bay of his B-52 and, after cutting it loose, riding it down to its Soviet target like a cowboy on a bull at the rodeo. Just before the climax of explosions, Dr. Strangelove emerges in triumph from his wheelchair to stand up and exclaim, "Mein Führer, I can walk." I have seen the film many times and Kubrick's ending always gives me goosebumps for its perfect fusion of insanity, male sexual aggression, cold war militarism, and what critic Robert P. Kolker calls Kubrick's dark warnings about "the eternal return of fascism" (Kolker 2017: 159). Kolker adds that "the revelation of real danger in the midst of manufactured terror is irresistible precisely because it is a revelation that only the viewer can recognize amid the bluster, madness, and impotence of the characters in the film" (2017: 179). "Like all great satire," Kolker concludes, "*Dr. Strangelove* requires a multilayered response of laughter at absurdity and a recognition of the deadly seriousness of its target" (206). As this conclusion suggests, *Dr. Strangelove* mixes dysphoric and euphoric elements. The film involves spectators in fear, paranoia, disgust, and the risk of apocalyptic death, all dysphoric elements, but uses grotesquerie, irony, and absurdity to elevate viewers above the action, encouraging them to laugh euphorically at the villains and fools careening toward catastrophe.

Male sexual fantasies, US militarism, and "the eternal return of fascism" were frequent themes in several of Kubrick's subsequent films, especially *2001: A Space Odyssey* (1968) and *Full Metal Jacket* (1987). *Jacket* returns to a realistic depiction of fascism and sexism in the US army and its deadly consequences for US soldiers in Vietnam. The film follows Private "Joker" and some of his US Marine buddies through basic training, where their Drill Instructor Sergeant baptizes them with new names that they carry with them throughout the film. Joker watches helplessly as overweight Private "Gomer Pyle"—the TV character name given him by the Sergeant to indicate his gullible innocence—is bullied and emasculated by their Drill Instructor. Pyle snaps and kills his tormenter, then himself, on the day before they will be shipped to Nam. Reduced to adolescent sexuality, Joker's platoon is met by a Vietnamese pimp and his prostitute, who do business with the boys, before the Marines wander into the battle for Hue during the Tet offensive of 1968. There they encounter a female Vietcong sniper who decimates several in their platoon by playing on their adolescent fantasies of altruistic heroism.

Images of and dialogue about "Mickey Mouse" ironically dot the film. "What is this Mickey-Mouse shit," demands the Sergeant, before Pyle kills him and himself in the shower room of the barracks. At the end of the film, the Marines sing "The Mickey Mouse Club Song," the theme from Disney's *Mickey Mouse Club* TV program in the 1950s, as they march off to their camp. As Kolker remarks, "The powerful marines are reduced to powerless children, playing games of life and death, their sexuality reduced to the banality of the infantilized impotent" (Kolker 2017: 177). Locking US troops into fantasies of adolescent sexuality and heroism during boot-camp and beyond would have long-term consequences for US politics. The last generation of males to face the draft, many Vietnam vets returned to the States bitter about their service abroad, cynical about democracy at home, and eager to kill enemies they viewed as anti-American. Many of them would become easy targets for authoritarian politicians.

Full Metal Jacket is a deeply dysphoric film. Kubrick keeps an ironic distance from his characters, nearly all of them White males, even when they are undergoing emotionally shattering experiences. Likewise, his characters nearly always keep an emotional distance from each other. Their primary exchanges, aside from occasional and finally mind-numbing small talk, have mostly to do with giving and receiving orders, punctuated occasionally by fatuous monologues and brief flashes of disbelief or anger. In short, life in the military has been mostly reduced to alienated hierarchical relationships, which play out in cold, bureaucratic spaces, or destroyed and confusing Vietnamese battlegrounds. Like *Dr. Strangelove*, *Jacket* broke sharply with filmic and televised conventions and depictions of military life in the 1950s and early '60s. Gone was the jokey banter among ethnically and regionally distinct platoon members from films about WWII, the delightful flim-flams of Sergeant Bilko on TV, and the "There is Nothing Like a Dame" and "New York, New York" numbers from Broadway and Hollywood musicals featuring soldiers and sailors at home and abroad. Kubrick's troops are mostly jerked up adolescents or alienated automatons, equally as ready to kill themselves as others.

6.2 From the Great Society to Reagan's Neoliberalism, 1960–1985

The first twenty years of the postwar era, 1945–1965, was a relative Golden Age for most White Americans. Economist John Kenneth Galbraith aptly named America *The Affluent Society* in 1958 for its widely shared material abundance. White labor union families, boosted by rising wages, acquired new homes in record numbers. To take another important indicator, electricity consumption tripled during the 1950s, its rise based primarily on the availability of inexpensive petroleum. Of course not all Americans shared equally in consumerist affluence. The average incomes of Blacks, Asians, and Hispanics lagged far behind those of Whites, and many Protestant Americans discriminated against Jews and Catholics as well.

Although these groups had made some progress compared to White Americans since the nineteen teens, African Americans were still denied substantial economic and political equality. This began to change when Lyndon Johnson assumed the presidency in 1963 after Kennedy's assassination and won a full term, together with a Democratic House and Senate, in 1964. A Texas New Dealer and a master of legislative politics, Johnson pushed to transform America into a "Great Society" and soon Congress passed laws to expand national medical insurance, federal housing, and higher education, plus civil and voting rights. Johnson correctly foresaw that the Voting Rights Act, which empowered federal examiners to register qualified minority voters in the South, would end Democratic power in the region for at least a generation. Major decisions by the Supreme Court in the mid-1960s also expanded the rights of poor and minority citizens. The Court held that states must provide attorneys to poor citizens at state expense, endorsed the principle of "one man, one vote" for congressional districts, struck down anti-miscegenation laws, and required police to advise suspects of their constitutional rights after arrest. Enthusiasm for Johnson's Great Society began to wane after 1966, however, as the President poured more money and troops into Vietnam. Then, in 1968, Johnson told a surprised public that he would not seek the presidency in the forthcoming election.

This decision proved to be the first of several shocking political events over the next six years. The racist Democratic Governor George Wallace, who had defied federal orders to end segregated education in Alabama, ran in several Democratic primaries in 1968 and won some of them. When anti-war protesters demonstrated against the war in Chicago at the Democratic National Convention in 1968, Mayor Daly ordered his city police to violently end the demonstrations, leading to a police riot and many injuries. After Nixon's election to the presidency, he ordered the bombing of neutral Cambodia in 1970 as a means, he said, of shortening the war in Vietnam, but this led to mass protests on several university campuses and the shooting of some students by the National Guard troops at Kent State and Jackson State. Following Nixon's reelection, investigators eventually traced a 1972 break-in at Democratic Party Headquarters in the Watergate Apartments back to the White House and the coverup of the burglary implicated President Nixon. After the Supreme Court ordered Nixon to release taped conversations about the coverup in 1974, Nixon resigned from office before he could be impeached by the House.

Although the presidency of Jimmy Carter was less eventful than Nixon's second term, it held dystopic surprises as well. From 1978 until 1980, double-digit inflation and soaring interest rates hit American consumers for the first time in the postwar era, causing significant financial disruption to many household budgets. Earlier in the decade, in 1973, Americans had stood in line at gas pumps for the first time ever, when OPEC, the Organization of Petroleum Exporting Countries, decided to temporarily stop the flow of oil from the Middle East to the West in order to jack up prices. Finally, one of the most stunning events of the late 1970s was the decision of the new revolutionary government in Iran to make hostages of US embassy employees in Tehran; the hostage crisis lasted over 400 days. Many of these events between 1965 and 1980 marked significant breaches of public trust and social cohesion. For most White Americans, these events were simply unthinkable during the postwar years of progress and prosperity. The apparent episodes of chaos during these fifteen years helped to hasten the triumph of neoliberalism.

In the midst of these surprising difficulties, a new elite group of ultraconservative economists, businessmen, and politicians—believers in and advocates for what has come to be called neoliberalism—challenged the waning power of traditional New Deal and conservative Republican assumptions. The "liberalism" in neoliberalism refers to the classical, laissez-faire liberalism of the mid-nineteenth century, when many

Victorians believed that economic relationships must be separated from governance so that the unimpeded "natural laws" of the marketplace could generate progress for humankind. After WWII, Austrian economist Friedrich Hayek convinced many corporate and political leaders that "prices and the market system, in human society, were the equivalent of natural selection in biological evolution," explains world historian Patrick Manning (Manning 2020: 225). Hayek's ideas about a natural, unregulated economy persuaded many US economists, among them Milton Friedman, whose Chicago School group already rejected the dominant Keynesian model of governmental regulation. In effect, Hayek and Friedman "adopted a timeless theory of biology at the very moment that biological theory was undergoing complex transformation," notes Manning (2020: 225). Backed by conservative business interests, neoliberalism prevailed among many economists, most Republicans, and some Democrats by the end of the 1970s. After Reagan's election in 1980, Republicans began decreasing regulations on business and protections for workers and the poor and increased tax cuts and other policies favoring global corporations, monopolistic mergers, and the rich. These policies and the ideology behind them decimated minority hopes for social and economic justice and substantially widened the gap between the super-rich and everyone else.

The rise to political power of a business-class elite in the 1970s had its roots in the rout of Goldwater Republicanism in the 1964 election and the political success of Johnson's Great Society programs left conservatives with little option but to reinvent themselves. One of the early architects of the conservative resurgence was Lewis Powell, the future Supreme Court justice, who began ringing alarm bells about a crisis facing American capitalism in 1971. In a memo invited by the American Chamber of Commerce entitled "Attack on American Free Enterprise System," notes journalist Jane Mayer, Powell "laid out a blueprint for a conservative takeover" (Mayer 2016: 89). The enemies of "free enterprise" were not merely the socialist and Black power radicals on the left, according to Powell, but mainstream institutions, including many Protestant churches, ivy league universities, the media, and most politicians. The key to changing this dominance, said Powell, was to capture public opinion by creating new institutions, such as academic programs, intellectual journals, and foundations, that could shape debates about public policy by pushing conservative and libertarian beliefs. As Mayer explains, Powell's widely circulated memo "electrified the Right;" prompted "a new breed of wealthy ultraconservatives to weaponize their philanthropic giving in order to fight a multifront war of influence over American political thought" (92).

Several new ultraconservative institutions proliferated in the 1970s, funded by millionaires and billionaires. Joseph Coors and Richard Mellon Scaife poured money into the Heritage Foundation and dominated its agenda. By the early 1980s, Heritage Foundation sponsors included Amoco, Boeing, Chevron, Dow Chemical, Exxon, General Motors, Mobil Oil, R. J. Reynolds, Union Carbide, and several other Fortune 500 companies. Charles Koch used his oil millions to establish the Cato Institute, which turned academics and journalists into flacks for libertarian ideology. With money from Scaife and others, political operative Paul Weyrich founded the American Legislative Exchange Council (ALEC), a group that funded ultraconservative politicians to take over state governments. By cofounding with Jerry Falwell the Moral Majority organization, Weyrich also enlarged the pro-corporate political fold by including religious conservatives. One of Weyrich's key targets was labor unions, understood by all ultraconservative groups as a major impediment to achieving their goals.

Prompted by Powell's initiative, the CEOs of General Motors, General Electric, Dupont, Citibank, and other major corporations founded the Business Roundtable in 1972 in order to take a more direct role in national politics. One of their first initiatives was to kill the Industrial Reorganization Act, proposed by Great Society Democrats, which was intended to reinforce anti-monopolistic practices. In alliance with Chicago School economists, Business Roundtable advocates argued that monopolies, through economies of scale, could actually save consumers money, a possibility that more than outweighed their power to set prices and undermine competition. The CEOs found some unlikely allies among consumer-rights advocate Ralph Nader and many new Democrats in Congress, dubbed "Watergate Babies" because they had been elected after Nixon's resignation, but knew more about US policy in Vietnam than about the long struggle against monopoly power in the twentieth century. Eager to overthrow the seniority system in Congress, the Watergate Babies blocked the Industrial Reorganization Act, overthrew key Democratic chairs in the House, and handed several victories to the neoliberal monopolists. According to economic historian Matt Stoller, Chicago School economists "convinced not just the right wing but the left that antitrust, and more broadly democracy, as practiced in the middle of the twentieth century, was not only inefficient, but countered the dictates of natural economic systems and science itself" (Stoller 2019: 369). One immediate beneficiary was Sam Walton, the owner of Walmart, whose price manipulations bankrupted thousands of retailers in the South and Midwest and made him the richest man in America by 1985.

Reagan's victory in the 1980 election not only legitimated a broadscale attack against unions and antitrust laws, but, more sweepingly, against the authority of government to regulate business at all. According to Mayer, "One measure of the movement's impact was that starting in 1973, and for successive decades afterward, the public's trust in government continually sank.... By the early 1980s, the reversal in public opinion was so significant that American's distrust of government for the first time surpassed its distrust of business" (Mayer 2016: 108). Mayer concludes, "By creating their own private idea factory, extreme donors had found a way to dominate American politics outside the parties" (2016: 110). Reagan and his advisors, largely in agreement with the neoliberals, understood the political revolution that had occurred. The new President passed out copies of the Heritage Foundation's playbook, which contained 1,270 policy proposals, to all members of Congress. A Foundation spokesman later noted that the Reagan administration passed 61 percent of them. These included huge tax cuts for wealthy individuals and corporations, abolishing the controls on oil and gas prices imposed by Nixon, and a further tax cut specifically aimed to enhance oil profits.

6.3 Dysphoric Films Before and After the Reversal, 1950–1980

As noted at the top of this chapter, Turchin's structural-demographic shift from an integrative to a disintegrative cycle occurred in the US between 1965 and 1980. Turchin did not deploy Whitehouse's coding of religious rituals to better understand the structural-demographic trend reversals that he found for the US polity during those years. But we can look at popular dramas on film and television through Whitehouse's lens to sharpen our sense of what this major transition meant for the dysphoric stories Americans were telling themselves both before and after the present "age of discord" set in. In particular, two dysphoric genres—horror and crime films—flourished in the US from the 1950s through the '80s, bridging the transition between Turchin's integrative and disintegrative phases of social cohesion.

For example, *Them!*, a 1954 film featuring giant, radioactive ants that threaten Los Angeles, and a 1962 episode from *The Untouchables* television series, which contrasts the incorruptible virtue of heroic federal agent Eliot Ness with a businessman who has ties to the mob, were popular examples of horror and crime dramas during the high point of Turchin's integrative phase. These may be usefully contrasted with two later dysphoric films from the same genres, *Night of the Living Dead* (1968) and *The Godfather* (1972), both of which gained enormous popularity during the start of the gradual fall of the US into social disintegration. *Night/Dead*, a protean example of later zombie movies, features an embattled group of survivors, all of whom eventually die, and the film maintains the fiction that more cannibalistic ghouls lurk in the darkness ready to terrorize and devour us. Likewise, Francis Ford Coppola's *The Godfather* marked a major turning point in the genre of mafia-centered crime dramas, with its evocation of sympathy for a mob family, cascade of ritual killings, and cynical view of governmental law and order.

This study has benefitted so far from my ability to combine Whitehouse's understanding of the social function of euphoric and dysphoric rituals with cognitive scientific insights into the foundations of narrative drama. But it has not yet been possible to explore the likelihood of a correlation between the rise of dysphoric dramas in modern, complex societies and widespread indicators of a decline in social cohesion, a correlation that Seshat's conclusions lead us to expect. Turchin's work in *Ages of Discord* provides the structural-demographic analysis that allows for this possible correlation to be substantiated, at least insofar as two pairs of dramatic films may begin to demonstrate. Will the dysphoric elements present in *Them!* and in the episode from *The Untouchables* spike upward in the later, post-downturn horror and crime dramas, *Night/Dead* and *The Godfather*? If they do, this suggests that the relationship Turchin and Whitehouse have already found correlating dysphoric rituals with a slide in the indicators of social cohesion also applies to popular dysphoric dramas on film and television.

The primary dysphoric element in *Them!* is the image of the ants themselves, explained in the film as mutants created by the atomic explosions at Alamogordo, New Mexico. Roughly six feet tall and ten feet long, they crawl out of holes in the ground and lurk in the storm sewers of Los Angeles on hairy, spindly legs with bloated bodies, to threaten humans by stinging them to death with formic acid or crushing their bodies with their lobster-claw-like mandibles. Disgust and fear of mutilation or death—all grounded in evolutionary realities and listed by Whitehouse as dysphoric emotions—provide the highpoints of the action. The climax of the plot occurs when the army is brought in to fight the ants in the storm sewers of L.A. A heroic, altruistic police sergeant rescues two boys in one of the sewer pipes, but is squeezed to death by an ant before he can escape. Finally, the army arrives with the star of the drama, a scientist who specializes in ants, who tells us that new queen ants have just hatched, but luckily have not yet left to start new nests. Flame throwers provide a fiery inferno that kills the beasts and fries the rest of their eggs. At the end of the film, however, the scientist warns of unknown dangers to come: "When Man entered the Atomic Age, he opened the door to a new world. What we may eventually find in that new world, nobody can predict" (Weisbart and Douglas 1954).

Despite this somber tone, which pervades the film and accompanied the likely audience response of fear and disgust in many scenes, *Them!* has several euphoric moments that occasionally take the edge off the elements of horror. The scientist is given some quirky, comic lines that raise a few smiles, and his good-looking daughter, another scientist, provides what Hollywood in the '50s understood as "sex appeal," which attracts some appreciative attention from a male FBI agent. Although their mutual attraction never amounts to a romantic subplot, it does provide occasional distraction from the general tone of anxiety. Then there are some short comic episodes with minor characters—a befuddled airplane pilot and a couple of alcoholics who would rather hit the bottle than talk about the giant ants they saw. Dysphoric elements predominate over euphoric ones, but that's to be expected in a horror sci-fi movie.

In this regard, the structure and point of view of the film ensure that it stays within the conventions of horror films and will be interpreted accordingly. Like many in this genre, the first third of the movie plays like a detective story, with the police and the FBI trying to figure out what is causing the murders and property destruction they uncover in the New Mexico desert. The giant ants are kept off-screen during this suspenseful buildup, although we do hear them occasionally, along with screams from their victims. The first ant the spectators see appears behind the scientist's daughter, alone in a pocket of the desert, oblivious to the danger and looking in the direction of the camera. Such conventional shots of innocence-in-danger (and soon rescued by her FBI boyfriend) occur often in *Them!*

For the category of social cohesion, Whitehouse lists examples such as oath-taking, the recognition of obligations among families, and actions that reflect the willing assumption of civic duty. The anthropologist recognizes that strong public norms generally accompany social cohesion in complex societies, while familial ties of altruism and/or honor often bind groups together in less complex ones. Although set against the background of possible nuclear war between two complex superpowers, *Them!* shines with civic trust and pride! Despite some minor tensions among public figures, the local police cooperate with the FBI, the scientist and his daughter work well with both of them, and when the mystery of the giant ants is solved, the federal government responds quickly to the threat, declares martial law in L.A., and sends in the army. In the parlance of the day, the authorities "contain" the threat from the nuclear ants, allowing innocent Americans to go on with their virtuous lives. Yes, the Atomic Age will kick up new threats in the future, but *Them!* also assumes that the scientists who brought us these problems, in alliance with other trustworthy public servants, also know how to protect us from "them."

This assumption is starkly at odds with the story world brought to life in *Night of the Living Dead*. The nightmarish plot of this thriller also builds on the legacy of "the Atomic Age," in this case the fear that "radioactive contamination" might cause the dead to rise from their graves and seek to eat the living. *Night/Dead* depicts cannibalism and intergenerational family murder and introduces an African American protagonist and racial conflict into its story—taboos that conventional Hollywood horror flicks had avoided before 1968. Lacking nearly any euphoric elements, it shows ravenous ghouls eating the charred remains of human flesh, and features a daughter, reanimated after her death, biting into her father's body and stabbing her mother to death with a masonry trowel.

The main plot is simply told. Ben, an energetic, responsible Black man in his 20s tries to help several White adults and teens save themselves from the zombies that are trying to penetrate an old farmhouse in western Pennsylvania where they are temporarily holed up, but his efforts are continually frustrated by their incompetence, fear, and racism. At one point, a frightened father tries to shoot Ben, but he wrestles the rifle from the older man and must kill the father to save himself. At the end of the film, Ben is the only non-ghoul character remaining in the farmhouse. Hearing gunfire in the morning, he emerges from the barricaded cellar to see a local sheriff and his vigilante deputies shooting the remaining zombies and signals to them that he is alive. But a deputy takes Ben for another zombie, shoots him in the head (the only way to kill zombies in the film), and throws him on the pile of ghouls to burn.

In his *Why Horror Seduces*, evolutionary cultural scholar Mathias Clasen argues that *Night/Dead's* zombies trigger defensive adaptation responses in their viewers related to human fears of predation and contagion: "We fear agents that have the will and the capacity to eat us and we have strongly aversive reactions to cues of contagion, such as the odor and sight of decomposing flesh" (Clasen 2017: 98). As Clasen notes, the ghouls of the film—the characters never actually call them "zombies"—are also weirdly uncanny. We all know that dead people should not be able to move around, much less try to eat the living. But these predatory and contagious monsters do both in *Night/Dead*.

George Romero, the writer and director of *Night/Dead*, borrowed some elements from previous Hollywood horror films, but scrambled them to purposefully confuse his audience and unsettle their conventional ways of interpreting his film. Instead of gradually building an atmosphere of mystery and tension, Romero begins by turning a simple outing involving a brother and sister's visit to a father's grave into an ambiguous situation with what may be the accidental death of the brother and the paranoic flight of the sister away from a stranger in the cemetery. Only later, when the sister is safe in the farmhouse, do we learn that she was right to flee from that flesh-eating ghoul. Soon after that, however, the sister sinks into a numbed stupor, and the character who might have provided the conventional innocent female victim of the horror film becomes mostly irrelevant to its plot until Romero decides to reanimate her later in the movie. For these and similar reasons, the usual interpretative paths for audiences to follow cannot be trusted in *Night/Dead*.

Nor can the conventional sources of morality and justice be relied upon to operate responsibly in the terrifying and grotesque world of the film. In contrast to *Them!*, families cannot be trusted to protect their own and the public authorities—the police, hospital officials, and federal government in *Night/Dead*—do not coordinate their efforts. When Ben is finally able to listen to news reports on radio and TV about what is happening and how to deal with the zombies, the coverage is haphazard and partial. The scientists, for example, are not sure why the dead are rising from their graves. Ben hears a report that a nearby hospital is a safe haven and, following that hopeful tip, gasses up his truck to take the wounded survivors there, but a gasoline spill upsets his plans and the torch he was using to protect himself from the ghouls starts a fire which leads to the truck's explosion. With families dysfunctional and public help impossible to reach, *Night/Dead* feeds on its own isolation and claustrophobia until Ben is the only human left alive. Then, when White law and order does arrive the next morning, it shoots Ben down like a criminal. Except for the racist actions of the police, all of the other ordering institutions and norms of social cohesion are in disarray. And White racism, of course, only serves to advance social disintegration.

Romero did not invent zombies—by 1968 such "living dead" characters had already been featured as mindless automatons in films about Haitian voodoo—but he did invent the subgenre of horror films that features the lurching, ravaging, cannibalistic monsters who fear fire but can only be killed by a bullet to the brain. *Night/Dead* continued to play in film houses for a decade after its release and it has been translated into more than 25 languages. Even today it ranks as one of the most popular downloadable films worldwide, in part because Romero neglected to copyright it. Hundreds of zombie films have been successfully produced and marketed since 1968 and film historians also credit *Night/Dead* with influencing the subgenres of "splatter" and "slasher" horror films.

Because crime dramas continued popular from the 1950s through the '70s, two films involving the mafia—an episode from *The Untouchables* (1962) and *The Godfather* (1972)—also offer insight into the sea change that occurred in indicators of social cohesion in the US during the 1960s and early '70s. *The Untouchables*, based on a fictionalized account of a Prohibition agent (Eliot Ness) and his team of Feds who fought crime in Chicago in the early 1930s, ran weekly on ABC Television for three seasons from 1959 to 1962. (Although produced by Desilu for television, *The Untouchables* was shot as a film.) I will examine a 1962 episode entitled "The Case Against Eliot Ness." Portrayed by Robert Stack, the award-winning star of the series, Ness accuses a businessman-promoter, Mitchell Grandin, of murder in a moment of anger at a public meeting organized to facilitate the Chicago World's Fair of 1933. The TV audience already knows that Grandin is in league with the mafia, which perpetrated the killing of the previous promoter so that Grandin and the mob could profit from running the Fair. Grandin sues Ness for libel and most of the 50-minute episode involves a search for three witnesses who can testify that Grandin was implicated in the initial murder. "Can Eliot

Ness protect at least one witness so that he can clear his name in court?" is the primary dramatic question of the show.

Like most films featuring mafioso bosses and murders before 1970, dysphoric elements predominate over euphoric ones. The boss in this episode is Frank Nitti, who (the audience is led to understand) took over Al Capone's gang when he was sent to jail for tax evasion. Just as Ness's upright behavior to find and protect the witnesses leads to admiration, Nitti's actions to wipe them out evoke disgust and disdain. The mob kills two of the three witnesses and it looks like Grandin might go free and win his case. Then Ness figures out a way to trap the two-timing wife of the last witness into flushing out her husband. Although this witness is killed in a shootout, Grandin does not know that he is dead and Ness induces him to plan to kill the already-dead witness in a darkened room. The film-noir trap works; Ness gets a photograph of Grandin, knife raised over the dead body of the witness, the promoter is booked, and "the case against Eliot Ness" is thrown out of court. A few grim jokes from the Feds and some boasting and glad-handing from Grandin provide occasional semi-euphoric relief from the tension.

The Hollywood conventions of mafioso crime films, set in the 1930s, changed little in mafia-centered episodes from *The Untouchables*. In response to criticism from Italian-American organizations that *The Untouchables* would perpetuate negative stereotypes of their ethnic group, Desilu Productions made some minor casting changes but continued to depict Italian-American gangsters as boastful, reckless, and bloodthirsty. "The Case Against Eliot Ness," like most mafia films, is a melodrama with unambiguous figures of good and evil. There can be little doubt that the primary TV audience for the episode knew how to interpret the story.

As in *Them!*, the Feds under Eliot Ness work cooperatively with other federal, state, and local agencies of law and order. Although the citizens of Chicago backing the World's Fair initially side with Grandin, Ness recognizes their good intentions and remains courteous and deferential to them. The patience and hard work of the Federal agents starkly contrast with the vainglorious recklessness of Grandin and the gangsters, who fight among themselves and fail to take advantage of their early success. Poetic justice is served at the end of the melodrama; the final shot shows Ness and his team passing through the entryway on the opening night of the World's Fair. For the 1962 television audience, this image connecting the incorruptible Feds to "The Century of Progress," as the 1933 Fair was named, linked trust in the federal government with the progress of mankind. What better way to validate the polity's protection of American social cohesion for the future!

The massive success of *The Godfather* just ten years later helped to undercut public confidence in the power of the government to guarantee justice and social trust in America. The highest-grossing film in 1972 (and for a time the highest grossing film ever made), *The Godfather* claimed ten Oscar nominations and was eventually ranked as the second-greatest film in US cinema history (just behind *Citizen Kane*) by the American Film Institute. Two *Godfather* sequels (in 1974 and 1990) followed and the new respect it won for the mafia helped to launch several other mafia-centered films and TV series, notably *The Sopranos*.

Unlike earlier mafioso crime dramas, the screenplay by novelist Mario Puzo and director Francis Ford Coppola pulls the audience into a family situation that mostly excludes questions of public justice. Instead of focusing on the Feds vs the mob, the film examines what happens when an extended mafia family, the Corleones, lose their patriarch and must find a new boss to run the family business and adapt it to modern challenges. In the course of these difficulties, Michael Corleone (played by Al Pacino), initially an outsider to the rackets run by his family, emerges as its reluctant leader and finally becomes its new and more ruthless godfather. Consequently, the Puzo-Coppola script required what Whitehouse understands as checks on spectator interpretation that were very different from those standardizing the performances and receptions of most mafia films before 1972. In brief, the rituals that gave authority to the role of godfather within the mafia family had to be underlined and the old and new Corleone godfathers had to be accorded sufficient sympathy, despite their obvious connections to murderous violence, to ensure that the audience would understand their legitimate and necessary role within the mafia family. Whatever one thinks of the public morality of the film and its effects on social cohesion in the US, *The Godfather* succeeds brilliantly in meeting both of these challenges.

The insular focus on mafia family values begins with the first scene of the film. In the midst of a joyful outside wedding celebration for his daughter, Don Vito Corleone (Marlon Brando) conducts business inside with Italian Americans who seek his advice and ruthless intervention. A father tells the story of the near rape of his daughter and asks the godfather for justice, for example. In another situation, a godson requests his godfather's help in securing a Hollywood contract so that he can become a movie star. The relationship between the petitioners and the godfather is essentially personal and feudal; Don Corleone avoids the legal complexities of modern justice to play judge, jury, and enforcer in each of these situations. Regarding the Hollywood contract, Don Vito dispatches his consigliere to make the film producer "an offer he can't refuse" and the producer gives in after discovering the severed, bloody head of his prize racehorse in his bed. What might have been interpreted as vengeful and authoritarian in earlier mafia-centered films is understood here as a nasty but necessary strategy to protect the power of the godfather and, by extension, his family.

With competition among the New York mafia families driving most of the plot, the Corleone family initially appears more humanitarian than the other gangs because Don Vito, unlike rival godfathers, does not want to involve his family in illegal drug trafficking. This rivalry leads to the shooting and eventual recovery of Don Vito and it begins the family's search for one of his sons to assume the mantle of the godfather's authority upon Don Vito's death. The long middle part of the film reveals the utter corruption of the police and the politicians in New York, who are profiting from the mafia's "business" ventures. Not surprisingly, none of the figures representing public law and order approach the squeaky-clean morality of Eliot Ness and his "untouchables." After Don Vito's fatal heart attack in 1955 (while the sympathetic old man is playing with his grandson in a flower garden), son Michael steps into his role. The new Don arranges to participate in the church baptism of a baby for whom he stands as godfather on the same day as his Corleone avengers' murder the other New York dons who had opposed his family.

The usual emphasis in previous mafia-centered films on dysphoric to the near exclusion of euphoric elements in the story also shifts in *The Godfather*. The initial wedding festivities include a wide range of positive, emotionally arousing scenes, with dancing, entertainment, toasts, feasting, and illicit sex in an upstairs bedroom. Hiding out in a safe village in Sicily to escape assassins eager to avenge his own revenge-killing against his family in New York, Michael falls in love with a beautiful local girl and marries her in an idyllic wedding ceremony. Still, dysphoric elements prevail and shock. In addition to the bleeding stallion's head in the producer's bed, there's the violent attempted murder of Don Vito, the knife driven through the hand and into the table of a Corleone capo caught pretending he's a rat, and Michael's murder of a crooked cop and two rival gang members for the attempt on his father's life. Then, in Sicily, Michael's new bride is killed when a car bomb, meant for him, blows her up instead. Back in the US, Sonny, Don Vito's first son and heir apparent to become the next godfather, savagely beats his sister's skirt-chasing, wife-thrashing husband and then is machine-gunned down in the blood-soaked toll-booth massacre scene.

Director Coppola saves the worst for last. At the climax of the film, the preparations for and killings of the Corleone enemies are intercut with solemn rituals baptizing the baby and investing the new godfather with divine authority in a lavish church ceremony. As Pauline Kael pointed out in her review of the film, Coppola followed the "uncoercive, 'open' approach to the movie frame" pioneered by Jean Renoir (Kael 1972: 7). This approach allows the spectator to "roam around in the images" (1972: 7) without the director's mandating specific interpretations of each scene. Again, Coppola presents a bloody action with ethical ambivalence. Is the church ritual meant as critique or simply as ironic counterpart to the revenge slayings? While both interpretations seem possible, neither finally matters. The sanctification of new godfathers and murderous vengeance are both made to seem necessary in the predatory world of the mafia. Many audience members may have responded with ambivalence, as well. Their evolved fears of violent death probably repulsed many, but drove others to yearn for psychological protection under the authoritarian wing of a mafia boss.

The Godfather sacrifices the social complexity and cohesion of the nation-state for feudal loyalties owed to an authoritarian family. Several scenes of the film, in addition to its overall plot, underline the supremacy of allegiance to one family and the necessity for a strong alpha male as its absolute leader. Scenes of oath-taking and ring-kissing bookend the film—first in fealty to Don Vito then to Don Michael—leaving no room for or even patience with the difficulties of democracy, the rule of law, or prying press discussions in the public sphere. The one voice in opposition to the drift of the film toward patriarchal authoritarianism is Kay Adams (Diane Keaton), who marries Michael after his return from Sicily. When midway through the story she asks Michael how he can justify the murders committed by his family, he answers that his father is "like a president" (Ruddy and Coppola 1972), even to the point of having people killed. In effect, the 1972 film invited spectators to choose between the war in Vietnam and a family that must kill to survive in the gang wars of New York. By this point in the action, most spectators have made their choice; never mind the racketeering, the murders, and the questions of justice that also weigh in the balance. At the end of her review,

Kael remarked, "When 'Americanism' was a form of cheerful bland official optimism, the gangster used to be destroyed at the end of the movie and our feelings resolved. Now the mood of the whole country has darkened, guiltily; nothing is resolved at the end of *The Godfather*, because the family business goes on" (1972: 9).

The two post-1967 dysphoric dramas, *Night of the Living Dead* and *The Godfather*, substantially depart from the generic conventions and interpretative checks of pre-1963 horror and crime dramas. The result is an increase in dysphoric shocks in the later dramas and a sharp decline in dramatic incidents that uphold social and political cohesion through prosocial emotions and affirmative public norms and actions. Instead, these later dramas represent the institutions of social cohesion above the level of the biological family—including the police, the federal government, broadcast journalism, public health, and the political process—as too corrupt to be reformed. In response to such a brutal and lawless society, *The Godfather* allows spectators to justify the patriarchal control and predatory murders practiced in one mafia family to ensure its survival. In some ways, *Night/Dead* goes even further. With the return of the dead to attack the living, the nuclear family is no longer even safe from its own children. This film, the original horror zombie movie, shows the gradual isolation of a man who attempts to practice the norms of traditional social cohesion only to be rejected by others in need of his help and finally killed because of the color of his skin.

Although there are no data that would allow us to trace the preferred politics of those audience members who were especially affected and ideologically influenced by *Night/Dead* and *Godfather*, we can combine Whitehouse's conclusions about the appeal of certain kinds of rituals with Tuschman's generalizations about the universality of the left–right political spectrum to support some informed speculations about the probable politics of these spectators. Recall that Whitehouse empirically linked the appeal of dysphoric rituals to those individuals who had undergone identity fusion, rather than group identification, as their primary mode of socialization. According to Whitehouse, such fused individuals would favor hierarchical rather than egalitarian modes of authority and they would support strong tribalistic affiliations involving ethnocentrism, religiosity, and sexual heteronormativity. Relating to their tribal group as an extension of their nuclear family, these individuals would also look for opportunities to sacrifice themselves altruistically for in-group tribal members.

As Tuschman's investigation of the relevant surveys reveals, these characteristics clearly line up on the conservative side of his universal liberal (left)-conservative (right) spectrum. In fact, some of these individuals might appropriately be located on the extreme right, authoritarian side of that continuum. In general, then, we can speculate that conservatively inclined individuals would be more likely than those on the liberal side of the divide to find the genres of horror and crime films appealing. It may be that Hollywood producers before 1965 correctly intuited the generally conservative orientation of these genres because they usually found room in such films to include a few euphoric elements that could also keep liberal spectators attentive and interested, as well as likely to return to their televisions and movie houses for more of the same. After 1965, however, when what might be loosely called the *zeitgeist* for dramatic narratives shifted to the Right in the wake of disintegrative historical trends, horror and crime films favoring a stronger dysphoric package apparently gained audiences and popularity. Consequently, it is likely that the embrace of tribalistic and authoritarian values in *Night/Dead* and *Godfather* did help to move those who were already inclined to embrace neoliberalism further toward it.

References

Clasen, Mathias. 2017. *Why Horror Seduces*. New York: Oxford University Press.

Dolan, Frederick M. 1994. *Allegories of America: Narratives, Metaphysics, Politics*. Ithaca: Cornell University Press.

Eckstein, G., and B. Kowalski. 1962. The Case Against Eliot Ness. In *The Untouchables*, A. Armer. Los Angeles, CA: ABC.

Kael, Pauline. 1972. Retrieved: 19 June 2019 from https://newyorker.com/magazine/1972/03/18/alchemy-pauline-kael.

Kennedy, Robert F. 1971. *Thirteen Days: A Memoir of the Cuban Missile Crisis*. New York: Norton.

Kolker, Robert P. 2017. *The Extraordinary Image: Orson Welles, Alfred Hitchcock, Stanley Kubrick and the Reimagining of Cinema*. New Brunswick, NJ: Rutgers University Press.

Kubrick, Stanley. 1964. *Dr. Strangelove: Or, How I Learned to Stop Worrying and Love the Bomb*. USA: Columbia.

———. 1987. *Full Metal Jacket*. USA: Warner Brothers.

Manning, Patrick. 2020. *A History of Humanity: The Evolution of the Human System*. Cambridge: Cambridge University Press.

Mayer, Jane. 2016. *Dark Money: The Hidden History of the Billionaires Behind the Rise of the Radical Right*. New York: Anchor Books.
Ruddy, A., and F. Coppola. 1972. *The Godfather*. USA: Paramount.
Stoller, Matt. 2019. *The 100-Year War Between Monopoly Power and Democracy*. New York: Simon & Schuster.
Streiner, R., and G. Romero. 1968. *Night of the Living Dead*. USA: Image Ten.
Turchin, Peter. 2016. *Ages of Discord: A Structural-Demographic Analysis of American History*. Chaplin, CT: Beresta Books.
Weisbart, D., and G. Douglas. 1954. *Them!* USA: Warner Bros.

CHAPTER 7

Neoliberalism and Political Realignment, 1980–2015

As noted in the last chapter, Turchin found that increasing numbers of the general population aspired to elite status and some of them moved into the elite during the 1965–1980 period. While some economists, sociologists, and others at the time greeted this change as a sign of the health of the American economy and proof of the actuality of the "American dream," Turchin, with his knowledge of past cycles of sociopolitical disintegration, understands that an expanding elite does not "raise all boats," the usual metaphor applied to justify the enormous disparities in wealth between the rich and the average. Instead, a growing elite usually foreshadows economic stasis, future difficulties for the general population, and trouble ahead for social cohesion. This is partly because nearly all elites throughout history have tried to increase their wealth and status *vis a vis* other elite groups. Their eyes are typically fixed on comparisons with others in their class and they tend to bask in ideologies such as neoliberalism that ignore the effects of their own scramble for position on those beneath them.

7.1 American Life Under Neoliberalism

In his magisterial *Age of Fracture*, an insightful complement to Turchin's *Ages of Discord*, cultural historian Daniel T. Rodgers notes that the myth of the market became one of the dominant concepts in American life during the 1970s:

> The term "market" that insinuated itself into more and more realms of social thought meant something much more modest than the financial markets' churning, and, at the same time, something much more universal and audacious. It stood for a way of thinking about society with a myriad of self-generated actions for its engine and optimization as its natural and spontaneous outcome. It was the analogue to Reagan's heroes in the balconies, a disaggregation of society and its troubling collective presence and demands into an array of consenting, voluntarily acting individual pieces. "You know, there really is something magical about the marketplace, when it's free to operate," Reagan told the nation in early 1982 as the motif of limitless dreams was swelling his speeches. (Rodgers 2011: 41)

As Rodgers points out, "magical" markets proliferated even as actual gasoline, financial, and stock markets veered wildly in the 1970s and early'80s. Instead, the market as a master metaphor for producing social good was a testament to neoliberalism's success in capturing the public's imagination.

By the 1990s, states Rodgers, "Faith in the wisdom and efficiency of markets, disdain for big government taxation, spending, and regulation, reverence for a globalized world of labor pools, free trade, and free floating capital [had become] the world's dominant economic ideology" (Rodgers 2011: 75). The Republicans trashed their traditional concerns about budget deficits in the rush to facilitate more tax cuts, even abandoning Republican President George H. W. Bush at reelection time for his failure to live up to his promise of "no new taxes." Many Democrats also jumped aboard the neoliberal bandwagon. Monopolistic mergers under Clinton actually surpassed those under Reagan and Bush; there were 166,310 such deals valued at 9.8 trillion during the eight years of his presidency, compared to 85,064 mergers valued at 3.5 trillion from 1981–1992. The Democrats under Clinton backed away from big government politics, finally settling for privatizing many governmental services and allowing tax credits for the working poor. The World Bank and the International Monetary Fund, which had helped nations weather market failures in the past, now loaned money that depended upon a country's promise to deregulate trade, end price subsidies, and relax restrictions on foreign investments. Not surprisingly, several political and economic commentators confused markets with democracy and consumers with citizens. According to Citibank's Walter Wriston in 1992, "Markets are voting machines; they function by taking referenda" (quoted in Rodgers 2011: 75). And journalist Thomas Friedman wrote in 2000 that markets "had turned the whole world into a parliamentary system [where] people vote every hour, every day" (75). Why bother with elections at all if the capitalistic marketplace could register the public will more regularly and efficiently than the ballot box?

Of course one reason to continue to hold elections was to mask the political reality that the US was becoming more of a neoliberal oligarchy than a democracy by 2000. Certainly, Koch, Scaife, and many of the wealthy ultraconservatives who controlled the Republican party by the turn of the century feared democracy and sought more power to shape the nation's economy through its politics. In *Dark Money*, Jane Mayer quotes several critics who, by 2016, were warning that the power of wealth is antithetical to the goals of a flourishing democracy. She also cites what she calls Thomas Piketty's "zeitgeist-shifting book" (2016: 15), *Capital in the Twenty-First Century*, which lays out the economic logic behind the rise of world oligarchs in the current neoliberal era. Piketty's dynamical analysis of the world economy demonstrates that the rich will continue to get richer in comparison with the general population simply because inherited wealth will continue to grow at a faster rate of return, unless governments intervene to reverse these inexorable trends. Left alone, the "magical markets" of capitalism guarantee that the wealth gap between the haves and the have-nots will widen even more, until the super-rich control more than half of the world's wealth. Piketty's recent *Capital and Ideology* (2020) does not deny this conclusion, but qualifies its causality, shifting much of the blame from economic dynamics to politics and ideology. His general solution to the problem of extreme inequality in the 2020 book is radical economic redistribution.

The increasing wealth gap and the rise of big money in politics tilted the political playing field after 1970 away from those social groups stuck near the bottom of the economic hierarchy in the US, including poor Whites, Hispanics, and nearly all other minority populations. During the Civil Rights period from the mid-1950s through 1968, African Americans had successfully pushed back against the overt restrictions of the Jim Crow era, which stripped Black citizens of voting rights, consigned many to economic destitution through sharecropping, convict leasing, and redlining, and enforced its mandates through brutal policing and occasional lynching. Despite divide-and-conquer efforts by the powerful to fragment the potential power of the poor, Black leaders and others formed alliances among several minority populations and progressive unions to push for reforms. The liberal narrative that emerged from the Civil Rights movement, energized by some success in Johnson's Great Society programs, was a heroic drama demonstrating that Blacks and Whites could work together to deliver political rights to all citizens. The rise of Black Power in the mid-'60s and the urban riots that erupted in the wake of the killing of Dr. Martin Luther King, Jr. in 1968 dented but did not destroy that optimistic narrative that looked forward to a future integrated America in which "We Shall Overcome" might attain the status of a second national anthem.

Rodgers' *Age of Fracture*, however, demonstrates that a different narrative, which ignored the realities of race, power, and poverty, displaced that heroic story. He begins by noting that pigmentation and other indelible marks of race—unlike the changeable signs of occupation, class, and other, more freely chosen identities—made those who bore them, "willingly or unwillingly, into bearers of memory and history" (Rodgers 2011: 113). Memory and history, however, played no role in neoliberal narratives centered on the market, where everyone was supposedly a free agent living outside of historical constraints, with the ability to choose among employment opportunities, consumer products, and social identities. In Reagan's America this fact alone shut Black people out of much of the national conversation. Nonetheless, Rodgers notes that polls of African Americans during the 1970s and 1980s suggest that most Black citizens did not retreat from the integrationist goals of the Civil Rights era; they continued to seek the same political rights and economic opportunities enjoyed by other groups. The primary political rhetoric of Jesse Jackson's "rainbow coalition," when he ran for President in 1984 and 1988, for example, was "our fair share," "parity," and "reciprocity."

Conservative power brokers, however, were inventing powerful strategies and narratives to deny the poor a fair share of the American dream. One was Nixon's Southern Strategy, which recognized that Republicans could capture most of the South for their party by playing on the same tribalist fears that segregationists had mobilized to keep the South nearly "solid" for Democratic racists from the 1890s to 1964. Another was the narrative of ethnic identity. One of the most significant storytelling events of the 1970s was Alex Haley's *Roots*, a run-away best seller as a novel in 1977 that was soon serialized on TV. The performance of *Roots* dramatized Haley's successful search for his ancestors and their struggle to survive the Middle Passage, endure the degradations of slavery, and eventually flourish in the US. Soon, White ethnic groups that had opposed busing and the integration of their urban neighborhoods were borrowing this heroic tale for their own purposes—celebrating their own families, their struggles, and their own ethnic "roots." The success of *The Godfather* among some Italian Americans was an early forerunner of this strategy.

Other borrowings and inversions followed. Several conservatives who had railed against MLK in the '60s fixed on a phrase from his March on Washington address—King's hope that one day children would "not be judged by the color of their skin but by the content of their character"—to argue that MLK had actually pushed for a "color-blind" America. Conservatives wielded this quote and its implied, out-of-context policy endorsement in the affirmative action debates of the 1980s, the primary battle line for racially based civil rights programs from the mid-1970s into the 1990s. Opponents of affirmative action sought to turn every interview for a job and application for a loan into an ahistorical "color-blind" situation that denied the power dynamics and past injustices of American racism. As Rodgers remarks, "'Color blind' was an aspiration, a lawyer's tactical maneuver, a linguistic hijacking of the opposition's rhetoric. Above all, it was a way of making the legacies of race disappear.... Justice was not achieved by attention to history; justice was achieved by transcending the past" (Rodgers 2011: 130, 132). In America, however, then as now, Black-and-White citizens bore markers of racial difference that kept them indelibly tied to history.

Perhaps the most consequential battle over racial "privilege" during the age of fracture focused on welfare reform, but actually concerned the structural morality of poverty and crime. In the late 1970s, some conservatives and libertarians began to question whether so-called "underclass" Americans—those who apparently lived off of welfare checks, used drugs, made illegitimate babies, and engaged in looting during riots—actually deserved the privileges and entitlements of regular citizenship. Had a multigenerational "culture of poverty" encouraged these mostly Black teens and young adults to drop out of normal society to perpetuate their immoral lifestyle? Others pointed out that the critics had put the question the wrong way around; it was more likely that the hopelessness of poverty, not an alien form of culture, had driven some young Blacks to drugs and crime. But the debate spiraled downward, sparked by occasional urban riots and egged on by stories from President Reagan about "welfare queens" and attack ads employed by George H. W. Bush's presidential campaign in 1988 featuring the African American criminal Willie Horton, whom his opponent had paroled. In this atmosphere of racialized fear, the national government mounted its much-touted War on Drugs as one answer to the "problem" of a Black underclass and many state and local governments passed minimum sentencing laws that kept minor drug offenders in prison for years. Determined to pull the plug on the "culture of poverty," conservative moralists began to advocate cutting off welfare payments to single mothers who had additional children and compelling young adults in poverty to seek their future in the marketplace, whose abstract workings many Americans now agreed meted out economic morality. President Clinton acceded to pressures from Democrats as well as Republicans to "end welfare as we know it" in 1996, bringing to a close a patchwork system that had endured several modifications since the 1950s. Rodgers laconically concludes that "the formal question of what each might owe to all [regarding the ethics of economic justice] was set aside..... [In the age of fracture] it was hoped that the poor might be made Americans once more" (Rodgers 2011: 209).

From a coevolutionary point of view, however, it is foolish to expect the poor to act like middle-class Americans. Nearly all of the moralists of the 1990s, Black as well as White, blamed the so-called "culture of poverty" on individual character flaws, rather than on what poverty does to human psychology and behavior. "We are powerfully biased to look through the specifics of a given situation as if they were a pane of glass and to explain behavior based on the characteristics of a person," states psychologist Keith Payne, adding, "This bias has been replicated so many times that researchers took to calling it the 'fundamental attribution error'" (Payne 2017: 61). Moreover, flipping the reason for bad behavior from "bad character" to "bad environment" simply backs the explanatory frame into another version of the nature vs nurture conflict, which tied the social sciences in knots for the last half of the twentieth century. Rather, says Payne, "the goal should be to comprehend how human nature has prepared us to respond in resource-rich and resource-poor environments" (2017: 63).

Evolution designed our species to follow two different strategies when living in environments of poverty or affluence. Regarding reproduction and survival into old age—the two evolutionary strategies that have mattered for the success of our species—populations living in poverty will tend to produce many babies early in life in the expectation of losing several of their children and dying young, while more affluent populations will usually have fewer children, raise them expecting most to survive, and live longer lives. This second strategy, now available to many more families on the planet than ever before, might be called "invest in the future." In contrast, Payne terms the first strategy "live fast, die young" (Payne 2017: 66). Studies focused on how soon young women reach puberty have confirmed that the start of menstruation varies among women raised in poor and chaotic family situations compared to those living in wealthier and more stable homes. Having more babies earlier in life is only one of the consequences of poverty and instability. "Live fast, die young" parents, psychological experiments confirm, will gamble more often, focus more on immediate gratification than long-term planning, and take risks with their own and other's lives that appear rash and unnecessary to more affluent adults. Given the social and economic structures of life in the US, most risky behavior drives poor people into even more debt and poverty, as well as drug problems, homelessness, and violence.

Spike Lee's 1989 film, *Do the Right Thing*, asks about the practicality and morality of violence. Long-simmering racial and ethnic tensions among several social groups—African Americans, Hispanics, city police, Korean Americans, and Italian Americans—finally boil over into a fight, a killing, and a riot in the Bedford-Stuyvesant neighborhood of Brooklyn, New York. Like the narrative of the film itself, its textual epilogue offers two quotations that present different points of view about what happened. The first, by Dr. King, concludes: "Violence as a way of achieving racial justice is both impractical and wrong.… Violence ends by defeating itself. It creates bitterness in the survivors and brutality in the destroyers." This is followed by a quotation from Malcolm X, which includes: "… I am not against using violence in self-defense. I don't even call it violence when it's self-defense, I call it intelligence" (Lee 1989).

One of Spike Lee's most popular and celebrated films, *Do the Right Thing* won several awards and established Lee as a major, if controversial filmmaker. Lee wrote, directed, produced, and took a major role in the film as Mookie, a pizza delivery man in his '20s who lives with his sister and is struggling to support his Hispanic wife and their infant son.

It is clear from the outset that the neoliberal economy of Bed-Sty has left Mookie to scramble for a precarious livelihood; harassed at work, he must also stay on good terms with his customers in the neighborhood. Given the film's timing and subject matter–racism, police violence, and incendiary rioting in the late 1980s—it was bound to incite sharply different opinions among viewers and critics. Some White reviewers even worried that the film might incite riots from Black audiences. Several questioned whether Spike Lee as Mookie, who decides to throw a trash can through the window of Sal's Pizzeria soon after the police drive away with the body of a teenage boy they have killed, really "did the right thing." Other critics suggested that Mookie's action saved the life of Sal and his two grown sons because the crowd might have turned on them in their rage against the police killing. It is not a stretch to read much of the critical response to *Right Thing* through the lens of the Bad Character vs Bad Environment conflict that dominated arguments about the "culture of poverty."

A coevolutionary take on the film, however, leads the critic to different conclusions. Although the difficult ambiguities of the movie and this brief summary of its climax suggest a film that is deeply dysphoric, *Right Thing* strikes several grace notes of humor and humanity for most of its two-hours traffic on the screen. Lee's film introduces us to many of the colorful and quirky folks of the Bed-Sty neighborhood, including a friendly old drunk, a symbolic grandmother to the community, a jiving DJ, a mentally disabled man, a Korean couple who own a local grocery store, and Radio Raheem, who carries a boombox everywhere to blast the rock music of Public Enemy. Several sweet and comic scenes mix it up with arguments over racism and White privilege. One scene that features representatives of various groups slinging colorful racist epithets at another group even mocks the foolishness and futility of such name-calling. Although the temperature is rising throughout the hot summer day and into the evening of the film, most of the characters look past the insults to accept this status quo as the necessary live-and-let-live price for social life in the neighborhood. Lee puts the audience in the midst of a "culture" in which "poverty" clearly constrains many activities and enables a few others, but the results are not entirely the "right" (or wrong) thing.

As a consequence, the characters who cannot accept this difficult status quo stand out from the rest. These include Buggin' Out, a friend of Mookie's who objects that Sal's "Wall of Fame" in his pizzeria features only Italian American celebrities but includes no African Americans. Buggin' Out also recruits Radio Raheem in his plan to boycott Sal's pizza after Sal demands that Radio turn off his boom box before he will serve him a slice. On the other side of the racial divide is Pino, Sal's racist older son, who holds the African Americans of the neighborhood in contempt and wants his father to sell the pizzeria and move out. Mookie, needing to keep peace in the neighborhood, attempts to moderate the hot-headed demands of both sides, but cannot budge Pino or Buggin' Out. A fight erupts when Buggin' and Radio enter the pizzeria just before closing to demand that Sal display pictures of African Americans on his Wall of Fame and Sal smashes Radio's boom box when he won't turn down the volume of "Fight the Power." The fight escalates from there, the police arrive and put Radio in a night-stick choke hold, which kills him as the neighbors look on. After the police leave, Mookie throws the trash can through the window and soon after that Smiley, a mentally disabled young man, starts a fire that burns down the pizzeria. Firemen arrive and turn their hoses on some of the teens who are fighting them and looting the store.

If he had simply wanted to inflame racial tensions in the US, Lee might have ended the film with the climactic fire and looting. But the film narrative moves to a final scene the next morning between Sal and Mookie. After an argument with his wife, Mookie returns to the burned-out pizzeria to find Sal in order to collect two weeks of back pay. Feeling betrayed by Mookie, Sal gives vent to his bitterness, but Mookie insists on payment and Sal does "the right thing." The cautious reconciliation between the two of them brings the film down from the murderous identity politics of the previous night to the realities of poverty and relative wealth in the neighborhood. The audience learns that Sal will collect insurance for the fire but also that Mookie must find another way to support his wife and child, at least in the short term. Lee underlines the fact that Mookie, shown to be carefully counting bills at the start of the film, needs the money.

Indeed, the film anticipates the many links between Black poverty and the rise of White capitalism that recent historians have explored and revealed. In his essay "Capitalism" for *The New York Times Magazine's* "1619 Project," Matthew Desmond notes that "in order to understand the brutality of American capitalism, you have to start on the plantation" (Desmond 2019: 30). Desmond draws on the sociology of Joel Rogers to detail the type of "low-road capitalism" that began to flourish in the American South after 1810. Although partly reformed by labor unions and New Deal and Great Society programs, low-road capitalism returned to immiserate many US workers after 1980: "In a capitalist society that goes low, wages are depressed as businesses compete over the price, not the quality of goods; so-called unskilled workers are typically incentivized through punishments, not promotions; inequality reigns and poverty spreads... [Consequently,] a larger share of working-age people [now] live in poverty than in any other nation belonging to the Organizations for Economic Cooperation and Development" (2019: 32). Desmond compares worker rights in the US to those of other advanced capitalist nations in 2019 and finds enormous disparities in unionization, in the enforced regulations covering temporary work, and in the protections that workers have from being summarily fired from their jobs. In that last category, for instance, the US ranked "dead last" out of 71 nations in 2019 (32). "Historians have pointed persuasively to the gnatty fields of Georgia and Alabama, to the cotton houses and slave auction blocks, as the birthplace of America's low-road approach to capitalism," Desmond concludes (32). Significantly, in 1989, Mookie could already be fired at any time from his pizza delivery job and he knows it.

In the end, the film does undermine the superficiality of identity politics; neither side of the racial divide gets a "Wall of Fame," which has burned down. Although few critics understood it in 1989, the final scene also suggests that the statements of Malcomb X and MLK are not necessarily in contradiction; both can be morally correct. Mookie used intelligent "violence" during the riot to prevent the killing of Sal and his sons, but the riot also created "bitterness" in many of the survivors and only increased the "brutality" of the police. Some liberal spectators—clearly the film's targeted audience—may also have noticed that following the statements by the two Black leaders, the epilogue ends with a photograph of MLK and Malcolm X shaking hands and smiling at each other.

From a contemporary perspective, a major problem with *Do the Right Thing* is how little attention Lee paid to the role of the police in the unfolding of his narrative. Until they intervene to break up the fight between the two neighborhood boys and the pizzeria owner and his sons, the police are a relatively benign presence in the neighborhood. Their actions in strangling Radio Raheem and then driving away with his body to prevent the crowd from turning on them, however, are both murderous and cowardly. Had such a situation occurred in a film today, investigators from the press, from Black Lives Matter, and perhaps from

City Hall would be looking for video evidence of what had occurred, demonstrators would probably march against police violence, and there would likely be renewed discussions about how to reform urban police departments. Instead, in 1989 before the widespread use of cell phones and no Internet, Lee accepts police violence as a given and focuses his attention on how Sal and Mookie will patch up their lives.

In terms of the "culture of poverty" debates of the 1980s and early 1990s, *Right Thing* demonstrates that the either/or perspective of "character vs. environment" is irrelevant. Both Mookie and Sal have plenty of "character;" both rise above the simplicities of that dichotomy and the identity politics of the late 1980s to do the "right thing" for each other. Lee's film opened up other conflicts, however, that are not so easily resolved. Both major characters remain trapped in low-road capitalism; Mookie must scrounge for another low-skill, low-wage job and Sal knows his new business would fail if he offered his next delivery boy better security and wages. And despite the grudging respect each has for the other at the end of the film, Mookie, still necessarily committed to a "live fast, die young" life, has no realistic incentive to think he could ever become a businessman in Bed-Sty and adopt an "invest in the future" perspective about survival. *Right Thing* suggests that Mookie and Sal will both survive in their different ways, but the realities of coevolution and US capitalism will likely drive them further apart, despite the moment of recognition and respect they share together at the end of the film.

7.2 Realignment and Polarized Politics

In *The Great Alignment* (2018), political scientist Alan Abramowitz traces the realignment of demographic political party loyalties from the mid-1960s to 2010. Broadly speaking, says Abramowitz, the major challenges of the 1960s and 1970s—civil rights reforms, protests against the war in Vietnam, the push for gender equality, changes in American families, and the decline of labor unions—created opportunities among Republicans to induce some groups of voters to abandon their traditional political allegiances and align themselves with the GOP. In addition, an influx of immigrants and higher birth rates among nonwhite citizens led to a steady increase in the nonwhite population and most of these new voters turned to the Democratic Party to advance their rights and interests. Through the mid-1960s, Roosevelt's New Deal coalition of Northern progressives, labor unions, the White South, and big-city ethnics remained a potent force behind most of Johnson's Great Society initiatives, but that Democratic coalition, says Abramowitz, fractured along racial, ideological, and cultural fissures. Polling results show that the racial divide was the most consequential of these divisions, largely "because of how racially conservative white voters have reacted to the growing racial and ethnic diversity of American society" (Abramowitz 2018: 8). Ideological and cultural party conflicts had mostly to do with disagreements over the proper size and role of the state and beliefs primarily centered on abortion, religious observance, same-sex marriage, and related hot-button issues. As a result, concludes Abramowitz, "Americans can [now] be sorted into two camps: those who view the past half-century's changes as mainly having positive effects on their lives and on American society and those who view the effects of these changes as mainly negative. Since the 1960s, Americans in the first group have increasingly come to support the Democrats, while those in the second group have increasingly come to support the Republicans" (12–13).

The "great alignment" of Abramowitz's title—the gradual march toward conflicting racial beliefs, ideological passions, and cultural norms under each party's banner that has polarized US politics—is unprecedented since the era of the Civil War. From the 1880s into the early 1990s, both parties were "big tent" coalitions, with a variety of interests and beliefs vying for power and influence. This situation typically ensured that presidents could work with a House or Senate controlled by the opposite party and lawmakers could cooperate "across the aisle" to pass important legislation. The rise of political polarization, however, has led to what Abramowitz terms "negative partisanship"—voters and politicians who harbor "strongly negative feelings toward the opposite party" (2018: 5). Abramowitz has plotted the rise of negative partisanship on what the well-regarded American National Election Study (ANES) has termed a "feeling thermometer" scale and found a significant increase in negative feelings toward the opposition party between the 2000 and 2012 elections.

Negative partisanship has increased political polarization at every level of governance. House Speaker Tip O'Neill's famous remark that "all politics is local" was broadly accurate in the 1980s, when he uttered it. As Abramowitz notes, however, negative partisanship made party loyalty so important after 2000 that O'Neill's adage now deserves to be turned on its head: "Today, all politics is national" (2018: 118). As a consequence, one party dominates in most state and local elections, fewer citizens shift from one party to the other to cast their votes, and the number of swing states in presidential elections is relatively small compared to elections before 2000. Most voters see stark ideological and cultural differences between their party and the opposition. In the words of Abramowitz, "The Democratic base is pro-government, pro-choice on abortion, and pro-gay marriage.... The Republican base is anti-government, anti-choice on abortion, and opposed to gay marriage" (117). As historian Katherine Stewart demonstrates in her *The Power Worshippers: Inside the Dangerous Rise of Religious Nationalism* (2020), many Christian Republicans are also drawn ideologically to religious authoritarianism. Stewart links the rhetoric of many Protestant and Catholic Republicans to the doctrine of the divine right of kings in the Bible, an orientation that would later provide nearly solid support for President Trump's authoritarianism among the Christian Right, despite his obvious moral failings.

In addition, Abramowitz examined the growth of "racial resentment" among White Republicans and compared it to White Democrats between the Regan era and the Obama era. Contemporary social scientists measure racial resentment by asking citizens to rate how strongly they agree or disagree with statements such as: "Over the past few years, blacks have gotten less than they deserve" and "It's really a matter of some people not trying hard enough; if blacks would only try harder they would be just as well off as whites" (2018: 129). Data from the ANES surveys between 1980 and 2008 showed that the percentage of White Republicans scoring at a high level of racial resentment jumped from 44 to 64% in those twenty-eight years. (During the same period high levels of racial resentment declined modestly among White Democrats.) Abramowitz concludes that by 2016 "racial resentment was pervasive among Republican voters – a situation that would make it much easier for a candidate whose message focused on white racial resentment to win the GOP presidential nomination" (131).

This voter and party realignment began to emerge in political coverage by the media in the late 1980s. Among the more neoliberal pieces of legislation passed during Reagan's presidency was the repeal of the Fairness Doctrine. Established in 1949 by the Federal Communications Commission, the Doctrine required a "reasonable balance" of perspectives on every issue broadcast to the public on radio and television. Because its repeal eliminated the obligation of licensed broadcasters to represent opposing views on their networks, the 1987 law opened the floodgates to conservative talk radio. Historian Jill Lepore reports that the number of talk radio stations mushroomed from 240 in 1987 to nearly 900 by

1992 (Lepore 2018: 704). Rush Limbaugh, one of the first talk radio hosts to establish a national reputation, began taking phone calls from supporters and broadcasting his venomous accusations in 1988. Talk-show anchor Alex Jones began his apocalyptic *The Final Edition* in 1996 with the allegation that the domestic terrorists who had killed 168 people the previous year by bombing a government building in Oklahoma City had been framed by the feds and he had evidence to prove it. Jones would later claim that Senator Barack Obama had conspired with others to fake his birth certificate and consequently could not run for the presidency, a baseless conspiracy claim that would later be repeated ad nauseum by the host-anchor of a reality TV show, Donald Trump.

Talk radio began the political fragmentation of news coverage that has now become a normal part of the public landscape in the US. Two partisan cable news channels, MSNBC and Fox News, started in 1996. Both followed the 24/7 model of CNN's news reporting, which captured many viewers during the Persian Gulf War of 1991. MSNBC generally sided with President Clinton's centrist policies while Fox News often backed the oppositional, no-compromise stands of many Republicans. Although Fox trailed both of the more liberal stations in viewership at the end of 1996, the impeachment investigations of President Clinton boosted its prime-time ratings by 400%.

Neoliberal Congressman Newt Gingrich applauded the bifurcation of news coverage in the 1990s. Impatient with get-along, go-along Republicans in the Democratically held House of Representatives in 1978, the newly elected Congressman told a group of College Republicans, "You're *fighting a war. It is a war for power....* What we really need are people who are willing to stand up in a slug-fest" (ital in original; Levitsky and Ziblatt 2018: 147). Gingrich initially got little traction among Republicans during Reagan's presidency, partly because Democratic Speaker Tip O'Neill often found it expedient to compromise with the popular President. But his rejection of mutual toleration and embrace of hardball tactics won him a loyal following, allowing Gingrich to become Minority Whip in 1989 and finally Speaker of the House in 1995. Trumpeting calls for political warfare—calls which led to strikingly different effects among viewers when carried on MSNBC and Fox News—Gingrich led the House to reject budget negotiations with President Clinton, causing government shutdowns in 1995 and 1996. The Congressman reached perhaps the low point of his Speakership when he insisted on impeachment for Clinton, even though the President's extramarital affair never came close to what most citizens assumed were "high crimes and misdemeanors." As Representative Barney Frank noted, Gingrich "transformed American politics from one in which people presume the good will of their opponents, even as they disagreed, into one in which people treated the people with whom they disagreed as bad and immoral" (quoted in Levitsky and Ziblatt 2018: 149).

Prior to Clinton's impeachment, Gingrich pushed through a neoliberal law that Clinton signed in 1996 that would have disastrous consequences for the regulation of the Internet, just beginning to emerge as a powerful engine of communication. Despite the fact that the Defense Department had developed the Internet with taxpayer money, the neoliberal utopians of Silicon Valley believed that it must be free of governmental interference. In 1994 Gingrich used his neoliberal think-tank, the Progress and Freedom Foundation, to bring together cyberutopians, Republican political operatives, and libertarian true believers to craft a "Magna Carta for the Information Age." Announcing cyberspace as "the latest American frontier," the Magna Carta proclaimed that governmental policy for "the knowledge age" should "focus on removing barriers to competition and massively deregulating the fast-growing telecommunications and computing industries" (quoted in Lepore 2018: 732). The bill that followed did just that. As historian Lepore concludes, "Clinton's approval of this startling piece of legislation would prove a lasting and terrible legacy of his presidency: it deregulated the communications industry, lifting virtually all of its New Deal antimonopoly provisions, allowing for the subsequent consolidation of media companies, and prohibiting regulation of the Internet with catastrophic consequences" (2018: 732–733).

Before the Internet lifted off, however, primetime television delivered a valentine to Clinton's centrist Democrats called *The West Wing*. Written primarily by Aaron Sorkin and starring Martin Sheen (as President Bartlett), the weekly series premiered in September of 1999 and ran for six seasons. According to media historian Janet McCabe, NBC promoted *West Wing* as an answer to "the deep cynicism pervading American political life;" the series emphasized the potentially ennobling purpose of politics "in the "post-Lewinsky era of political scandal and partisan vitriol" (McCabe 2013: 3). McCabe adds that the series soon found its creative niche "by evoking a parallel political universe as America lurched rightward with a Republican administration under George W. Bush" (2013: 4). Liberal spectators disgusted or simply disappointed with

Bush's presidency could retreat to a fantasy White House each week in which witty and well-educated political professionals balanced calls of conscience against tough political realities as they struggled to gain victories for an idealistic President and his administration.

The primary target for this exercise in what McCabe, borrowing the lingo of television producers, calls "quality TV," was an upscale demographic that would give the network leverage with advertisers: "NBC executives never failed to mention how the political drama delivered a disproportionately high number of affluent, upwardly mobile, better-educated, urban-minded professionals, ranking first among primetime shows in the percentage of eighteen- to forty-nine-year-olds boasting a median income of $75,000 and above" (2013: 24). Although ratings for *West Wing* peaked in the 2001–2002 season, the series continued to deliver a more elite audience to its advertisers than any other primetime show. McCabe notes that Sorkin's dialogue even went so far as to occasionally compliment viewers for their discriminating intelligence. In one episode, at the end of complex negotiations, President Bartlett admits that his attempt to get an international treaty on arms control probably lost because 82% of the general public could not possibly "be expected to reach an informed opinion" on it. "Yet," suggests McCabe, "the implied message was that those watching *The West Wing* can.... [T]he imagined audience for the series was implicitly positioned," adds McCabe, "as that elite 18 percent able to follow intricate political argument" (25).

Many liberal viewers took the bait. *West Wing* provided a weekly shot of idealism in the midst of despair over Republicans' stealing the presidential election of 2000, the multiple tragedies of 9/11, the unwarranted crackdown on civil liberties evident in the Patriot Act in 2001, and Bush and Cheney's cynical war for oil with the invasion of Iraq. *West Wing*, however, restricted the arena of possible political action to such a small sphere of elite advisors and its law-professor President that the raucous, specifically accented voices and multicultural faces of popular democracy were rarely heard or seen directly on the show. Although the series was full of democratic "issues"—women's rights, the structures of racism, even (occasionally) wealth and income inequality—the audience mostly heard common folks expressing and/or protesting about these problems and concerns on a television monitor in the midst of a West Wing office, if it heard them at all, and filtered through the savvy remarks of the political professionals in the room. To include these faces and voices in any meaningful way—to feature DC inhabitants similar to Spike Lee's Brooklyn neighborhood, for example,—would have violated the well-made, tightly orchestrated dramaturgy of Sorkin's imagined world.

In his critique of monopolistic power, historian Matt Stoller remarks that *West Wing*, which was heavily influenced by Clinton insiders, "inspired a younger generation to embrace the ideals of the Watergate Baby generation, the closeness to business, the lack of willingness to assert public power" (Stoller 2019: 431). As Stoller explains, the Watergate Baby generation of Democrats was a group of idealists who believed that Washington politics could be fixed on the inside from the top down and were happy to ignore racism and economic inequality when they raised political donations among the corporate elite. By excluding "We the people" from significant political action on *West Wing*, the show primarily endorsed the naïve idealism of the Watergate Babies. The show's embrace of hopeful rhetoric and bipartisan politics occurred just a few years after the Gingrich revolution, which had already hardened right-wing resolve and pushed many Republicans toward polarized positions. Despite its democratic rhetoric, *The West Wing* demonstrated that national political conversations were best conducted by elites inside the Beltway, ironically confirming a later Republican complaint that national governance under the Democrats was isolated from the very people it was meant to serve.

7.3 The Better Angels of Our Nature

In contrast, the 2012 film *Lincoln* demonstrated that US Presidents, even in the midst of the polarized politics of a civil war, should risk alienating some of their supporters if the ultimate goal is the eventual social cohesion of the nation. Hollywood provided a film aimed at reversing the antidemocratic course of current US history with *Lincoln*, directed and produced by Steven Spielberg, starring Daniel Day-Lewis in the title role, and written by Tony Kushner. A predominately Center-Left film with a rough balance between euphoric and dysphoric elements, *Lincoln* is a political melodrama about the successful campaign in the waning weeks of the Civil War to pass the Thirteenth Amendment to the Constitution, which abolished slavery throughout the nation and effectively began the postwar process of Reconstruction. *Lincoln* enjoyed widespread popular success and was later seen by numerous high school students after it was freely distributed to over 37,000 school districts.

The film makes it clear that Lincoln had come to realize that slavery must be legally abolished before the war ended in order to guarantee a democratic future for the soon-to-be reunited nation. The Emancipation Proclamation had abolished slavery in the rebelling states and Lincoln knew that the Supreme Court might simply recognize the Proclamation as a wartime measure that had no efficacy after hostilities ended. Further, Lincoln understood that abolition and preserving the Union would remain a single cause only as long as the war continued, but there could be no guarantee of postwar abolition if the South sued for peace and agreed to rejoin the Union in exchange for the preservation of slavery within their states. The North, after all, had gone to war to repel Southern aggression and to prevent the spread of slavery into new states and territories; very few Northerners supported abolition at the start of the war. And Kushner's script makes it clear that many White Northerners in 1865 feared that full freedom for all African Americans would undermine their historical privileges and endanger their own political future. Lincoln and his political lieutenants knew they would have to convince, cajole, and buy off many Northern Democrats who were supporting the war but had little reason to support legal rights for Black citizens if sacrificing abolition could bring an end to the bloodshed. As in Kushner's stage success, *Angels in America*, *Lincoln* centers on ethics, minority rights, political power, and the future of US democracy.

Probably recognizing that his primary plot could have little directly to do with the immense efforts of Black Americans to liberate their race, Kushner opened the film with a brief episode dramatizing the climax of the Battle of Petersburg, which showed Black soldiers in the midst of bloody, hand-to-hand combat, killing and being killed. Voice-over narration during the battle also notes that the Confederates typically slaughtered captured African American soldiers rather than treating them as prisoners of war. In the next brief scene, the script has President Lincoln listening to Black soldiers, who tell him about their lower pay and poorer conditions in the Union army compared to the Whites. The script also found ways of including contemporary photographs of the realities of slavery and several conversations between White and Black characters in the White House and elsewhere about their fervent desire for freedom. The end of the film includes a scene between the Radical Republican Thaddeus Stevens and his Black housekeeper/common-law wife as they sit in bed together and she reads aloud the language of the amendment that he had helped to pass. Necessarily, however, Kushner could not have injected more Black agency into his narrative without undercutting its focus on the political struggle among Whites to pass the Amendment.

Kushner also emphasized the primacy of Northern racism and White supremacy that motivated opposition to the amendment. Even in January of 1865 with the war nearly over, conservative Republicans preferred to pursue peace with the South rather than accept what their leader calls the "abolitionist nonsense" of full freedom for former slaves. Because he needs their votes for the Amendment, however, Lincoln must accede to the conservatives' attempt to pursue peace. Several Northern Democrats openly express racist disdain for the Blacks who have been fighting and dying for the Northern cause. Their leader states that the Thirteenth Amendment will "niggerate America," even claiming that it will "manacle the limbs of white Americans," enslaving them to African Americans (Spielberg, 2012). In a short conversation about how to handle the Reconstruction to come, Stevens urges Lincoln to deal severely with the South because "white people cannot bear sharing the country's wealth with Negroes," a perspective borne out by the action of the drama. Kushner aimed his attacks on historical racism straight at contemporary America. By 2012, Trump's birtherism crusade and the Republican use of racist dog whistles had been attacking President Obama along similar lines for several years. *Lincoln* simply underlines the perduring traditions of White American racism that culminated in but certainly did not end with the Civil War.

Tellingly, Kushner's Lincoln adopts several successful strategies against the racists. First, he uses his power of appointment to reward a few wavering Democrats with modest governmental positions for changing their votes. Second, he talks with his potential allies among the Radical Republicans to get them to back off from their claim that Black people are the natural equals of Whites so as to prevent a vote on the Thirteenth Amendment along the lines of ideological purity, which would cause it to fail. Third, Lincoln also speaks privately to a few Democrats and creates an ethical space in which they might safely reconsider the effects of their racism. In speaking with George Yeaman, a Democrat from Kentucky who doubts that Blacks are smart enough to vote responsibly, for example, Lincoln acknowledges that he cannot foretell the future of Black political participation, but asks only that Yeaman "disenthrall" himself from the "Slave Power" and think again. As it turns out, Yeaman's vote for the Amendment begins a shift among a few House Democrats that finally ensures Lincoln's victory.

Finally, Lincoln outmaneuvers his opponents. The dramatic climax of the film occurs when the House Democrats urge that the Amendment be tabled and convince Republican conservatives to go along with them because they have heard that a Confederate committee is in the city seeking to negotiate a peace deal with the Federal government and they believe that the President has kept this information from them. At the time, a Confederate delegation headed by Vice President Alexander Stephens was waiting at the port of Hampton Roads for peace negotiations to begin, but Lincoln was able to write a note to the Democrats (and to his wavering Republican allies) reassuring them, truthfully, that no Confederate delegation was actually "in the city" of Washington to negotiate peace. When read to the House, Lincoln's equivocation allows the voting to proceed and at the end of the process, the Thirteenth Amendment passes by two votes.

Lincoln's insights, feelings, and outgoing humor are at the center of the action and drew a variety of responses from empathic spectators. Not surprisingly, Daniel Day-Lewis won an Oscar for his work and high praise from many historians, one of whom noted that Day-Lewis even caught the way the President walked. It is clear at the start of the film that the war has taken a terrible toll on Lincoln's health and happiness. His wife's continuing grief and guilt for the death of their younger son continues to wear on him as well. The ongoing realities of death and dismemberment are never far from Lincoln's imagination, brought home in one scene by his older son, Robert, who follows a soldier pushing a covered wheelbarrow up a ramp near a hospital and watches him dump its bloody contents onto the growing mound of amputated legs and arms. (Not even Branagh's *Henry V* can top *Lincoln* for the number of body parts discussed and displayed.) Nonetheless, Lincoln still finds time to tell humorous stories, which euphorically amuse and relieve the tensions felt by his listeners, both on the screen and in the auditorium.

With the passage of the Amendment, church bells ring out over Washington, soldiers and Black citizens celebrate, Lincoln hugs his young son Tad, and Stevens goes home to his wife. But the moment of euphoric celebration is short-lived. In an expert bit of dramatic writing, Kushner decided to announce the assassination of President Lincoln at a performance of *Alladin and the Wonderful Lamp*, enjoyed by Tad Lincoln and an accompanying aide from a side box in a very different theater from the one most spectators were surely expecting. When Tad clutches the railing of the box and breaks into tears, we willingly experience his loss as our own through empathy; the shock connects the audience with the historical immediacy of the nation's tragedy and grief. Kushner ends the film with a flashback to the famous ending of Lincoln's second inaugural address, "With malice toward none; with charity for all; with firmness in the right, as God gives us to see the right; let us strive on to finish the work we are in." The speech, of course, foreshadows Lincoln's vision for Reconstruction and leaves the audience of the film with the recognition that there is still much work to be done.

With the political quest for full equality as its driving action and major theme, *Lincoln* remains an inspirational film. In terms of Tuschman's left–right spectrum, the film is more radical than most Americans probably expected. The forces of cooperation and "charity for all" have triumphed over those of competition and racist division. Cross-racial xenophilia, though tenuous and based on unequal status, wins out over exclusionary tribalism. Regarding the tension between equality and hierarchy, *Lincoln* recognizes some modes of hierarchy as legitimate and necessary, especially in political and military affairs, but elevates coalition-building and egalitarian procedures and condemns the implicit authoritarianism of slavery. Finally, the film acknowledges the rough-and-tumble realities of democratic politics, cautions against the trap of ideological purity, and implicitly advocates for progressive reforms and constitutional amendments that will redress the failings of the first Reconstruction.

As historian Eric Foner understood in *Reconstruction: America's Unfinished Revolution: 1865–1877* (1988), the Radical Republicans attempted to rebuild the country as a racially egalitarian nation, but failed, due to financial corruption, racism in the North as well as the South, and a vindictive President Andrew Johnson, who undercut its initial reforms after Lincoln's assassination. Central to their efforts were the 13th, 14th, and 15th Amendments, which were designed to guarantee the rights of freedom, citizenship, and voting to the former slaves. Despite mounting difficulties, nearly two thousand Black men had held public offices by 1877, when Reconstruction was abandoned and politicians in the North turned their backs on vengeful Southerners, who began reclaiming political power through campaigns of repression and terror. By 1890, Southern racists, under the banner of state's rights, had turned most of the former states of the Confederacy into apartheid regimes in which White males (joined after 1920 by White females) could oppress all African Americans. So thoroughly did Jim Crow laws and the norms of a racial caste system erase the realities of cooperation during Reconstruction and the oppression that followed that many Americans who grew up during the twentieth century never understood the authoritarian state systems they helped to sustain. Foner's recent book, *The Second Founding: How the Civil War and Reconstruction Remade the Constitution* (2019), reclaims the legacy of the Reconstruction Amendments in an attempt to realize their promise of equal citizenship in a democracy for all.

Spielberg's and Kushner's *Lincoln* may be understood as a part of this project. Because the film relies on an activist President for its model of political effectiveness, however, the 2012 release of *Lincoln* was already too late to make an immediate difference in US politics. President Obama had only the first two years of his presidency during which Democratic leadership in domestic affairs was possible. After that, the polarized politics of the nation closed down the strategy of "working across the aisle," the key to Lincoln's effectiveness in the film of his name. Republican Mitch McConnell took over leadership of the Senate in 2010 after Obama passed relief bills and health care in his first two years, rendering a Lincolnesque presidency impossible for Obama through 2016. In addition, by 2012, the Internet had already transformed the nation in ways that made Lincoln's mix of heightened oratory, down-home humor, and political bribery quaintly out of reach.

7.4 The Internet and Neoliberalism

By the turn of the century, the Internet was beginning to undermine some norms and assumptions about politics and elections, but it would take the Democratic loss to Trump's Internet campaign in the election of 2016 for the analysts to begin to grasp the major political implications of this new medium. The Internet is more than a new communication medium; it also involves a crucial expansion of neoliberal capitalism. For this reason, Marshall T. Poe's summary of the structural implications of the Internet, especially as it compares to the audiovisual media of the last century, is instructive but limiting. Recall that Poe's structural approach looks at the major "attributes" of each medium and the networks they flourish within, then follows this causal chain to "social practices" and "cultural values" to track the historical effects of the medium examined. This approach is initially useful in understanding the continuing promise of Internet communication for individual and small group agents, but it neglects the realities of neoliberal capitalism.

Nonetheless, it is useful to begin with Poe's attributes of accessibility, velocity, range, and volume, all of which distinguish Internet communication from previous media regimes. Regarding accessibility, audiovisual and Internet media are both easily available to consumers—it is not difficult to use a television or I-phone—but they differ substantially on the matter of accessibility to producers. Compared to TV and film production, producing for the Internet is relatively accessible to amateurs; almost anyone can send pictures of their pets or children. The velocity of most Internet communication, especially when done by amateurs, is also very fast. The Internet tends to be dialogic, as well; you show me a short video of your favorite politician and ten minutes later I'll send you a video of my favorite musician. Regarding range, the extensive reach of audiovisual media encouraged its producers to consider (and pander to) the diversity of their audiences. Because of the enormity of the Internet, however, and the tendency of dialogic conversations and shared stories to carve out smaller groups of people who share specific conversations, Internet use in the US tended initially to encourage niche rather than mass audiences.

In terms of volume, the kinds of networks created by both audiovisual and Internet media are high, especially when compared to all earlier media of communication. Given the unconstrained nature of Internet production and flow, however, it is clear that there can be no real comparison between the two media in terms of sheer volume. In 2006, the average home in the US received 118 television channels. By 2015, however, a viewer could troll the Internet to find and watch nearly all of the reruns and many of the live events that were on TV in 2006, plus millions of other televised programs and old films. The Internet is now the world's media library and its capacity is without channels or walls. Such enormity is also a burden, however. Too many choices tend to be overwhelming, leading Internet users to radically curtail their searches and to remain with what they have come to rely on as entertaining websites and reliable sources.

How do these medium and network attributes play out in terms of social practices and cultural values? According to Poe, these four attributes are moving us closer to social practices that are "equalized," "democratized," "diversified," and "amateurized" (Poe 2011: 249). In turn, these practices, says Poe, should reinforce the cultural values of "egalitarianism," "deliberativism," "pluralism," and "individualism" (2011: 249). Although the Internet should, in general, be making us and much of the rest of the world more equal and less tribalistic, the opposite seems true. Since Internet communication among the millions took hold around 2005, nearly all indicators tell us that these generally democratic values and practices around the world have declined. Why the obvious discrepancy?

Poe published his *History of Communications* in 2011, before it became evident that privacy and democracy would become major problems on the Internet. In her path-breaking book, *The Culture of Connectivity: A Critical History of Social Media* (2013), Jose van Dijck begins by noting that "it is a common fallacy... to think of [social media] platforms as merely facilitating networking activities; instead, the construction of platforms and social practices is mutually constitutive" (van Dijck 2013: 6). As she explains, online interactions on Facebook, YouTube, Instagram, Twitter, and similar platforms are not directly interactive; they are controlled and structured automatically by algorithms. Van Dijck's book presents strong evidence that these algorithms, coupled with the massive popularity of social media, are reshaping the norms of sociality in many cultures of the world. Instead of the usual divide between public and private sociality, which began in hunter-gatherer bands and became institutionalized when humans settled down in villages and cities to maintain agriculturally based cultures, the usually clear boundaries between these two spheres are now blurred. Where there used to be fairly evident distinctions between how you acted in your home (including your bedroom) and what you could do in public places (such as courtrooms and institutions of worship), these divisions are no longer clear. The norms of social media networking now encourage users to respond to their "friends" anywhere they can get online, regardless of the situation.

According to van Dijck, popular social media are in the process of moving humanity from traditional and formal notions of community-oriented connectedness to the apparent informality of corporation-mediated connectivity. She prefers the term "connective media" to social media because the age of the Internet is shifting us out of mostly direct social interactions and into relations of "friending" and "liking" that are structured for profit by Facebook and other corporations. Most of the enthusiasts of social media have already accepted these new norms and—even if they understand their implications—they are not much bothered by the shift in some areas of their lives from traditionally "social" to "connective" relations. "For many ordinary users, it is difficult to recognize how Facebook [and other connective media] actively steers and curates connections. Moreover, it is far from transparent how Facebook and other platforms utilize their data to influence traffic and monetize engineered streams of information," says van Dijck (2013: 12). Unlike its enthusiasts, van Dijk understands connective media as a cultural-cognitive ecosystem that provides significant constraints as well as evident satisfactions.

Van Dijck seeks to move the discussion out of the older public/private paradigm for thinking about such issues and into the territory of cultural history and shifting norms. As she insists, "[N]ew norms for sociality and values of connectivity are not the outcome but the very *stakes* in the battle to conquer the vast new territory of connective media and cultivate its fertile grounds" (*ital.* in original, 2013: 20). Van Dijck understands the importance of normative behavior in coevolutionary terms. She knows, as well, that such momentous changes in social relations do not occur easily or quickly.

Further, she locates the emergence of these new norms in the ascendance of neoliberalism. Since the 1980s, national and global neoliberals, in addition to pushing for market deregulation, a reduction in taxes to support only necessary government services, and the relaxation of restraints on foreign investments, have maintained that the Internet can monitor and control itself. The culture of connectivity, says van Dijck, "is staked on neoliberal economic principles.... Platform tactics such as the popularity principle [e.g., "liking"] and ranking mechanisms [e.g., "trending"]... are firmly rooted in an ideology that values hierarchy, competition, and a winner-takes-all mindset" (2013: 21). Not surprisingly, van Dijck finds evidence that Google and the other dot-coms generally align themselves with other neoliberals in an attempt to blur or reset the boundaries among private, corporate, and public domains. Poe may be correct about the Internet's initial potential for orienting human cultures toward deliberative practices and egalitarian values, but the marriage of neoliberal capitalism and connective media algorithms is moving the Internet's ecosystem toward an authoritarian political culture.

Harvard Business School professor Shoshana Zuboff calls this international plundering of personal data for profit "surveillance capitalism" (Zuboff 2019: January 25). With a nod to Jason Moore's *Capitalism in the Web of Life*, Zuboff states: "Capitalism has always evolved by claiming things that exist outside the market and bringing them into the market for sale and purchase. This is how we turned 'making a living' into 'labor' and 'nature' into 'real estate.' Surveillance now claims private human experience as free raw material for translation into behavioral predictions that are bought and sold in a new kind of private marketplace" (2019:

January 25). As Zuboff notes, this innovation began twenty years ago when "Google discovered that the 'data exhaust' clogging their servers could be combined with powerful analytic capabilities to produce reliable predictions of user behavior.... The young company's ability to turn these surplus data into click-through predictions became the basis for a lucrative sales process known as ad targeting" (January 29). Surveillance capitalism markets accurate predictions of individual consumption habits based on its cheap access to the patterns of our online appetites and desires. Put another way, the Internet now mines a fifth "Cheap" for capitalist profit: the individual mind. Because surveillance capitalism and democratic governance are foundationally incompatible, Zuboff recommends that democratic world leaders ban the selling of such surveillance around the globe, just as the industrializing world banned slavery in the nineteenth century.

In this context, it is important to remember that the major US Internet industries are monopolies. Like the railroads and Standard Oil before them, the entrepreneurs of Google, Amazon, and Facebook built their economic power through price-fixing, buy-outs, and mergers, all made legal in the 1980s and 1990s through the neoliberal revolution. The key regulatory change for Google came in 1996, when a Federal Communications Commission report noted the possibility that the dominance of a single search engine could "push competitors out of the way," but decided to encourage self-regulation instead of competition (Stoller 2019: 446). In building his empire, however, Jeff Bezos had an inherent advantage denied to the competition because, notes Stoller in *Goliath*, "Google knew what its users were thinking" (447). Between 2004 and 2014, Google spent over $23 billion buying up 145 companies. Stoller concludes, "Today, with Google, Amazon, Facebook, we find ourselves in America, and globally, with perhaps the most radical centralization of the power of global communication that has ever existed in history" (449).

One result of the inducements offered by this expansion of capitalism into our minds, says Zuboff, is "a pervasive amalgam of [consumer] empowerment inextricably layered with diminishment" (Zuboff quoted in Silverman, ibid). A 2019 summary by psychologist Doreen Dodgen-Magee that draws on many studies underlines several of the troubling "diminishments" Americans now suffer from their Internet use. Put together in pre-COVID-19 times, the trends enumerated in this mega-study probably undercount the immersion and effects of online use on citizens. The average adult in the US spends more than eleven hours a day online. Teens live in the digital world about nine hours per day and even children between the ages of 2 and 8 average two-and-a-half hours daily. Further, Dodgen-Magee writes that "both a correlational and causal relationship between tech use and various mental health conditions has been established" (Dodgen-Magee 2019: March 19). These conditions include "higher rates of depression and anxiety among young adults who engage in many social media platforms than those who engage in only two," "poor cognitive functioning" from those who frequently multitask online, and "decreased well-being" in young adults who engage in Facebook, Snapchat, and Instagram use (2019: March 19). As a result of US Internet use, writes Dodgen-Magee, "Our attention spans are short, our ability to focus on one task at a time is impaired. And our boredom tolerance is nil. We now rely on the same devices that drive so much of our anxiety and alienation for both stimulation and soothing. While, for many of us, these changes will never roam into the domain of addiction, for others they have" (March 19). Dodgen-Magee concludes that "mental health professionals must begin to educate themselves about the digital pools in which their clients swim" (March 19).

7.5 Internet News and Politics

One of those pools is the news media. Although newspaper circulation has been in decline since the 1960s, around 2000 an increasing number of readers discovered that they could get the news "for free" online and the slow move from printed to televised news turned into a stampede to abandon both for the Internet. As historian Lepore notes, however, "The Internet didn't leave seekers of news 'free.' It left them brutally constrained. It accelerated the transmission of information, but the selection of that information – the engine that searched for it – was controlled by the biggest unregulated monopoly in the history of American business. Google went public in 2004. By 2016, it controlled nearly 90 percent of the market" (Lepore 2018: 737).

Lepore agrees with van Dijck and Zuboff that the rise of the Internet has further eroded the distinction between the public and private spheres. Print culture and the enlightenment had bequeathed to modern democracies a fairly clear distinction, as well as a dialectical give-and-take, between the views of a political elite and the "common sense" of the people—a public sphere that continued in the US, despite momentous changes, into the 1960s. Post-1945 audiovisual culture, however, was already bending the deliberative and democratic possibilities of this public political dialogue toward an embrace of celebrity and consumption. "With the Internet," notes Lepore, "that model yielded to a model of citizenship driven by the hyper-individualism of blogging, posting, and tweeting, artifacts of the new culture of narcissism" (Lepore 2018: 738). By mining online data, connective media companies could feed individuals only the news they wanted, leaving them in an echo chamber with others who agreed with them, a process that invariably radicalized their perspectives. Consequently, says Lepore, social media "exacerbated the political isolation of ordinary Americans while strengthening polarization on both the left and the right, automating identity politics, and contributing, at the same time, to a distant, vague, and impotent model of political engagement" (738).

The "impotent model" of politicking Lepore has in mind is the belief that expressing virtuous outrage on the Internet is, by itself, sufficient to change the political behavior of others. This model is impotent because the echo chamber effects of Internet outrage rarely, if ever, transcend the niche political cultures created by corporate algorithms to engage with opponents or even with those undecideds who might be open to change. Nonetheless, many citizens also used the Internet in old-fashioned ways to drum up support for direct, face-to-face involvement in political action. Such action began to make a difference in the wake of the global economic collapse of 2007–2008, caused directly by the Republican deregulation of real estate financing, but exacerbated by the concentration of monopolistic financial power in the private sector. The result was a dramatic fall in stock values by 20% between October of 2007 and the end of 2008. Over nine million Americans lost their homes in foreclosures, wiping out the savings of many middle-class workers.

The economic collapse also helped to elect Democratic Senator Barack Obama as President. Even though Obama, like Clinton and Bush before him, embraced neoliberal financial orthodoxy and bailed out bankers but not homeowners, the Republicans launched an Internet campaign soon after his inauguration to oppose many of his plans, especially his economic and health care reforms. Called the Tea Party, in memory of the patriots in Boston who had opposed British tyranny on the eve of the Revolutionary War, this opposition began on Fox News and quickly spread through Facebook, YouTube, and other connective media. Although stoked by anchors Glenn Beck and Sean Hannity on Fox, the Tea Party movement soon attracted funding from the Koch brothers, with the Heritage Foundation, the Cato Institute, and other neoliberals providing speakers, press releases, transportation, and Internet support. Used to operating behind the scenes at the level of policymaking, the Kochs and their libertarian friends began to understand how the Internet could channel their money into direct political action.

At the other end of the political spectrum, the Occupy movement began organizing through connective media in 2011 to protest income inequality and Obama's bailout for Wall Street. By 2011, the top one percent of US families were earning 24% of the nation's income, a percentage that had more than doubled since the late 1970s. With this fact in mind, the Occupy movement adopted the slogan, "We are the 99%." More anarchist than socialist, the Occupy movement gradually fell apart, but it did elevate the only socialist Senator in Congress, Bernie Sanders, to prominence. Sanders' outrage about poverty in the US sparked Internet conversations that ignited the Occupy movement, elevated him to celebrity status among young radicals, and would sustain the most successful grassroots, online fund-raising effort ever attempted in a Democratic primary before 2018.

Through videos posted on the Internet, Black Lives Matter (BLM) began protesting the police treatment of African American citizens in 2012. Founded by Black activists Alicia Garza, Patrisse Cullors, and Opal Tometi, BLM has publicly protested a range of injustices, from rape and violations of gay and lesbian rights to the mass incarceration of African American males and the structural reasons for Black poverty in the US. In this regard, BLM has called attention to the reimposition of segregated communities and schools that have left many Blacks in pools of poverty and oppression. BLM leaders are aware that Black progress toward racial equality stalled in the 1970s and took another hit in the Great Recession, when many Black families lost their homes, the chief source of wealth for middle-class Americans. Historian Isabel Wilkerson would later interpret these problems as a matter of "caste" in her 2020 book of that title. Whereas many Americans see racism as an individual failing, Wilkerson understands caste as social structure. It is "the granting or withholding of respect, status, honor, attention, privileges, resources, benefits of the doubt, and human kindness to someone on the basis of their perceived rank or standing in the hierarchy" (Wilkerson 2020: 70). From her perspective, the Civil Rights movement and the legal reforms it generated in the 1960s addressed overt acts of racism, but barely touched on the enduring realities of caste, which have continued to shape the social norms of many White citizens. BLM's chief branding strategy has been to use demonstrable police oppression caught on video by on-the-spot citizens with iphones as examples of normative police behavior to demand fundamental reforms in the administration of justice across the US.

In 2014, witnesses with smartphones photographed the killing of Michael Brown by police in Ferguson, Missouri, the shooting of twelve-year-old Tamir Rice as he played with a toy gun in a city park in Cleveland, and Philando Castile in Minnesota, as he was trying to tell the police about the licensed handgun he carried in his glove compartment. Many of these videos circulated to millions on the Internet, made local and national television news, and evoked substantial sympathy for the victims and their families and condemnation of the police. Brought to trial, however, the police who had committed the killings usually claimed that they felt threatened by the Black person they shot and juries acquitted nearly all of them of manslaughter. BLM worked with city governments to require that all police wear video cameras and turn them on, but such requirements, where passed, were seldom widely enforced. Under pressure from police unions and citizen groups that wanted tighter law enforcement, many local governments backed away from these and similar requirements, even though the Obama administration included such measures in their outreach programs to improve police–community relations. Nonetheless, providing video of police brutality remained an important strategy for BLM and its supporters and would provide crucial evidence in the murder of George Floyd by four policemen in 2020, touching off nationwide protests that challenged and began to transform some of the norms of policing.

Although Black Lives Matter could not claim to have reformed the institutional framework for resolving Black–White conflicts in policing situations by 2015, their videos and protests did challenge the traditional narrative of such events for some Americans. This challenge was tellingly apparent in the recent dramatic adaptation of the story of the Central Park Five, a series entitled *When They See Us*, broadcast on Netflix in 2019. The four-part series reframes the 1989 story of the five young Black and Latino boys who were railroaded by the NY police into confessing to the beating and rape of Trisha Meili while she was jogging at night in Central Park. Viewers follow the five boys as they venture into the Park for some fun on a summer night, get rounded up the next day by the cops and are pressured to construct "confessions" to enable city prosecutors to mollify the public outcry, go to trial, are convicted on falsified evidence, then (for four of them) sent to juvenile detention centers, and finally released. The final episode focuses on the fifth boy, Korey Wise, a sixteen-year old who was convicted as an adult and sent to prison, where he spends many days in solitary to avoid beatings and rapes by older inmates before (in an outcome that would defy dramatic credibility were it not true) meeting the man who actually raped Meili and was willing to confess to the crime. The epilogue puts all of the actors, young and old, who played the boys at different ages on stage with the actual Central Park Five, now thirty years older, for a discussion moderated by Oprah Winfrey.

Television critic Emily Nussbaum wrote an excellent review of *When They See Us* for *The New Yorker*, which underlines its major theme and cognitive-emotional appeal: "[I]ts main concern… is empathy. Not a syrupy, manipulative empathy but a rigorous one, meant as a corrective. As the title indicates, it takes boys who were seen as a group – reduced to an indistinguishable pack of animals – and insists that they be viewed as individuals, children worthy of love, and then, years later, men worthy of justice. If they're free, we're free" (Nussbaum 2019: 80). Black Lives Matter also used video to insist that each of the victims of normative police brutality they photographed be accorded the right to empathy and individuality, even if their legal rights and the rights of their survivors might not gain justice in a court of law.

Invariably, BLM and *When They See Us* continue to be viewed by most Americans through a left-right political lens. To be effective, all communication media must work in conjunction with the cultural possibilities of their times and the disintegrating norms of civility in America have opened up a widening rift in the racial divide of American life as well as undreamed of opportunities for racist demagogues. How different the historical rollout of the Internet and its political effects might have been in better times, when the forces of cultural integration and economic opportunity were on the upswing. As several commentators have noted, perhaps the closest parallel to the effects of the new medium of communication on present US culture is the period of the Protestant Reformation, when the success of the printing press helped to foment a hundred and fifty years of warfare. In this regard, it is probably too soon to judge the long-term "social practices" and "cultural values" that Poe attributes to Internet communication. He predicted social practices that would be "Democratized" and "Diversified," among others, and cultural values that included "Egalitarianism" and "Pluralism" (Poe 2011: 249). While it is obvious that we are a long way from these values and practices, it is also true that Poe's methodology cannot account for the historical context of neoliberalism. Change the context and some of these more enlightened, cooperative, and potentially universal ethics might flourish. The Internet revolution is still unfolding.

REFERENCES

Abramowitz, Alan. 2018. *The Great Alignment: Race, Party Transformation, and the Rise of Donald Trump*. New Haven: Yale University Press.

Desmond, Matthew. 2019. Capitalism. The 1619 Project, ed. J. Silverstein. *The New York Times Magazine*, August 18, pp. 30–40.

Dodgen-Magee, D. 2019. We Need to Take Tech Addiction Seriously. *The Washington Post*, March 19.

Foner, Eric. 1988. *Reconstruction: America's Unfinished Revolution, 1863–1877*. New York: Harper and Row.

———. 2019. *The Second Founding: How the Civil War and Reconstruction Remade the Constitution*. New York: W. W. Norton.

Lee, Spike. 1989. Do the Right Thing. United States: Universal Pictures.

Lepore, Jill. 2018. *These Truths: A History of the United States*. New York: W. W. Norton.

Levitsky, Steven, and Daniel Ziblatt. 2018. *How Democracies Die*. New York: Crown.

Mayer, Jane. 2016. *Dark Money: The Hidden History of the Billionaires Behind the Rise of the Radical Right*. New York: Anchor Books.

McCabe, Janet. 2013. *The West Wing*. Detroit, MI: Wayne State University Press.

Nussbaum, E. 2019. Prestige Trauma: Powerful Truths on "When They See Us" and "Chernobyl." *The New Yorker*, June 24, pp. 80–81.

Payne, Keith. 2017. *The Broken Ladder: How Inequality Affects the Way We Think, Live, and Die*. New York: Viking.

Poe, Marshall T. 2011. *A History of Communications: Media and Society from the Evolution of Speech to the Internet*. Cambridge: Cambridge University Press.

Rodgers, Daniel T. 2011. *Age of Fracture*. Cambridge, MA: Belknap Press.

Stewart, Katherine. 2020. *The Power Worshippers: Inside the Dangerous Rise of Religious Nationalism*. London: Bloomsbury.

Stoller, Matt. 2019. *The 100-Year War Between Monopoly Power and Democracy*. New York: Simon & Schuster.

Turchin, Peter. 2016. *Ages of Discord: A Structural-Demographic Analysis of American History*. Chaplin, CT: Beresta Books.

Van Dijck, Jose. 2013. *The Culture of Connectivity: A Critical History of Social Media*. Oxford: Oxford University Press.

Wilkerson, Isabel. 2020. *Caste: The Origins of Our Discontents*. New York: Random House.
Zuboff, S. 2019. Curb the Excesses of Surveillance Capitalism. *The Washington Post*, January 25.

CHAPTER 8

Neoliberal Politics, Polarized Films, and Authoritarianism, 2000–2020

In his 2016 "Preface" to *Ages of Discord*, Peter Turchin remarks that "the American polity today has a lot in common with the Antebellum America of the 1850s; with *Ancien Régime* France on the eve of the French Revolution; with Stuart England during the 1630s; and innumerable other historical societies that went through integrative/disintegrative cycles" (Turchin 2016: iv). I will draw primarily upon Steven Levitsky and Daniel Ziblatt's *How Democracies Die* (2018) to trace the downward spiral of US politics into the kind of polarization that corroded social cohesion and put the nation on the road to authoritarianism after 2000. In addition to chronicling the rise of Trumpism, I will look at the polarization of Hollywood films that helped to elevate Trump to the presidency—dramatic films that both reflected the disintegrative influences noted in previous chapters and encouraged US citizens to expect, accept, and normalize political extremism.

8.1 Polarization and Authoritarianism

Comparativist political historians Levitsky and Ziblatt have found the same patterns of democratic failure and the rise of authoritarian leaders in several of the recent polities of Europe and Latin America that now inflect the path of the US. "If one thing is clear from studying [democratic]

© The Author(s), under exclusive license to Springer Nature Switzerland AG 2021
B. McConachie, *Drama, Politics, and Evolution*,
Cognitive Studies in Literature and Performance,
https://doi.org/10.1007/978-3-030-81377-2_8

breakdowns throughout history," they state, "it's that extreme polarization can kill democracies" (Levitsky and Ziblatt 2018: 9). US citizens during the cold war understood that insurgents and/or the CIA might quickly topple democracies around the world through a coup d'état. Levitsky and Ziblatt, however, caution that most democracies these days erode slowly; elected leaders, political parties, and impatient publics are more likely to cause democratic breakdowns than generals, revolutionaries, and covert ops. "Donald Trump's surprise victory [in 2016]," they state, "was made possible not only by public disaffection but also by the Republican Party's failure to keep an extremist demagogue within its own ranks from gaining the nomination" (2018: 8).

As the authors explain, the primary problem with Trump as President was his "dubious allegiance to democratic norms" (Levitsky and Ziblatt 2018: 8). Historically,

> Democracies work best – and survive longer – where constitutions are reinforced by unwritten democratic norms. Two basic norms have preserved America's checks and balances in ways we have come to take for granted: mutual toleration, or the understanding that competing parties accept one another as legitimate rivals, and forbearance, or the idea that politicians should exercise restraint in deploying their institutional prerogatives. These two norms undergirded American democracy for most of the twentieth century. (2018: 8–9)

How Democracies Die demonstrates that partisan polarization has been the primary cause for the erosion of these two norms in the US, especially among the Republicans. "There are, therefore, reasons for alarm," conclude Levitsky and Ziblatt: "Not only did Americans elect a demagogue in 2016, but we did so at a time when the norms that once protected our democracy were already coming unmoored" (2018: 9). In addition to suggesting strategies to protect US democracy, *How Democracies Die* relates the recent political history that has backed the US into this dangerous decline. They agree with Abramowitz that the "great alignment" of both political parties along the lines of ethnic segregation, cultural loyalty, and ideological purity after 1970 played a significant role in the polarization that concerns them.

As we have already seen with the rise of Newt Gingrich to power in the 1990s, the norms of mutual toleration and forbearance first began to change inside the Republican Party. After Gingrich, the next Republican

Speaker of the House, Tom "the Hammer" DeLay, devised a pay-to-play scheme with Washington lobbyists that rewarded them with legislation if they agreed to fund Republicans for Congress. Democrats in the Senate began responding in kind; during the last two years of G. W. Bush's presidency, they blocked the confirmation of several judicial nominations through 139 filibusters, even though it had been an unwritten custom of the Senate never to deploy the filibuster against nominees to the federal bench. Following Obama's election to the presidency in 2008, Republican Minority Leader in the Senate, Mitch McConnell, vowed to use the filibuster and any other means necessary to block the President's nominees and programs in order to keep him "a one-term president" (2018: 162). McConnell orchestrated a record-breaking 385 filibusters to carry out this breach of forbearance between 2007 and 2012. Roughly a year before the end of Obama's second term, the Senate, under McConnell's leadership, refused to consider his nominee for the Supreme Court, Merrick Garland, effectively denying the right of the President to fulfill one of his constitutional duties. Such a refusal to exercise senatorial forbearance had not happened for a hundred and fifty years. McConnell would continue to ignore the norms of forbearance in his service to Trump's presidency.

Like Reagan, Donald Trump benefitted politically from his popularity on television, but TV had changed substantially from the days when the broadcast culture of prestigious anchormen elevated genial actors to the status of celebrity presidents. As James Poniewozik notes in his acclaimed *Audience of One: Donald Trump, Television, and the Fracturing of America* (2019), the rise of "rich and famous" braggarts, antiheroes like Tony Soprano, survival of the fittest competitions, and similar trends on TV created a neoliberal niche in which a narcissistic self-promoter could hide his serial business failings behind a façade of billionaire success and become the arbiter of "winners" on *The Apprentice*. The producer of the show, Mark Burnett, shared Trump's Hobbesian understanding of reality and his contempt for "losers." Each episode of *The Apprentice* ended in a setting called the boardroom, where Trump, in the role of a capitalist Godfather, rewarded the head of the competitive team he judged most likely to succeed and fired the other. *The Apprentice* premiered in 2004, but its initially high ratings declined steadily after the first year. Nonetheless, Trump hosted the show on and off for fourteen seasons. In 2007, NBC rebooted the show as *The Celebrity Apprentice*, with a glitzier set, less famous celebrities in supporting roles, and Trump as a more domineering shill for success. Many Americans apparently believed that playing the role of an authoritarian businessman on *The Apprentice* prepared him to govern effectively as the President of the US.

During the first few months of the Republican primary campaigns for the presidential nomination in 2016, *The New York Times* retained historians, psychologists, and political scientists to analyze all of Trump's public utterances in an attempt to understand his growing popularity with Republican voters. These experts concluded that his rhetoric of "vilifying groups" and "stoking [the] insecurities of his audiences" echoed "some of the [worst] demagogues of the past century," including Joseph McCarthy and George Wallace (quoted in Dean and Altemeyer 2020: 27). In contrast to these predecessors, however, the *Times* found that Trump was a more "energetic and charismatic speaker who can be entertaining and ingratiating" (2020: 27). As a consequence, the newspaper warned that Trump might be both more successful and more dangerous than previous US demagogues.

Apart from his take-charge persona and his racist rhetoric, Trump's major political advantage in 2016 may have been his amoral willingness to break several norms of forbearance and mutual toleration in the pursuit of high office. He knowingly encouraged Russia to turn many of his casual supporters into Internet bots-for-Trump and to dissuade key groups of Democrats from voting. In 2014, billionaire Robert Mercer, with the help of Republican strategist Steve Bannon, positioned UK-based Cambridge Analytica to help the Republicans identify "psychographic" profiles of potential Trump supporters by collecting data on 270,000 Facebook members through a personality quiz app. By mining Facebook for these users' "friends," Analytica built a database of 87 million voters who could be swayed to support the Republican candidate. Bannon passed this information on to the Trump campaign and Facebook's algorithms helped to create an informational bubble with its connective media that, with its feeds and filters, kept these Facebook users loyal to Trump and encouraged them to spread his propaganda. A 2018 study by scholars associated with Microsoft Research found that voting patterns in the election strongly correlated with the average daily fraction of users visiting websites that peddled Republican deceptions. In addition, the campaign ginned up Trump's connective media base of support in the four months before the election by spending nearly seventy million for ads on Facebook.

The Russian operation, which included several fake accounts on Facebook, Twitter, and YouTube, was already underway in the US by the time of Trump's nomination. "Black Matters US," set up to attract African American voters, for example, had accounts on the above platforms as well as Google+ , Tumblr, and PayPal. Such multiple platforming enabled the Russians to link posts among many of them, solicit donations, organize real-world rallies, and direct their viewers to a website, which they also controlled. On Election Day, Democrats watching the polls in several of the major cities of Pennsylvania, Michigan, and Wisconsin, where Trump's margin of victory was very small, reported decreased turnout in many wards with large Black populations. The Russians also hacked Democratic computers and stole information about their election strategies, which they shared with the Trump campaign. In later congressional testimony, Facebook President Mark Zuckerberg would admit that 126 million of his users may have seen content produced and circulated by the Russians before the 2016 election. At the start of 2018, the Justice Department would indict thirteen Russian nationals for operating social media accounts "for purposes of interfering with the U.S. political system" and "supporting the presidential campaign of then-candidate Donald J. Trump" (quoted in Lepore 2018: 780–781).

Although many election watchers initially assumed that Trump had successfully appealed to economically marginalized White workers, multiple studies later confirmed that "Trump's voters were motivated more by partisanship, nostalgia, and race-based anxiety than by a sense of being 'left behind' by a changing economy" (Reid 2019: 25). This is journalist Joy-Ann Reid's conclusion and she backs it up with substantial evidence. Reid cites a four-year study in the *Proceedings of the National Academy of Sciences* published in 2018 that Trump voters essentially believed that they had lost the privilege of being White in America. Another significant empirical study found that of the five "main drivers" of voting for Trump, three of them had to do with tribalism and the perception that the economy was structured to work against them. In brief, these voters feared the rising power of non-White ethnicities and thought that immigrants were taking "their" jobs (2019: 27–28). Reid also quotes a 2017 *Psychology Today* report that identified four traits of the typical Trump voter on the basis of extensive interviews: Authoritarian personality syndrome, a preference for alpha male dominance, "sparse contact with racial and social groups outside one's own," and a sense of "being deprived of something to which one believes they are entitled"

(29). According to Reid, ethnocentric tribalism and hierarchical entitlement played the primary roles in ensuring Trump's 2016 victory, with a preference for authoritarian leaders close behind.

Reid's analysis followed most of the methods and conclusions of mainstream social science, but slighted the changing understanding of authoritarianism. In brief, political psychologist Bob Altemeyer had been reworking and testing possible definitions of authoritarianism since 1981 and had concluded that attempting to separate the characteristics of authoritarian personality types from preferences for extreme tribalism or racism, rigid hierarchy, and alpha male dominance was the wrong approach; all of these characteristics (and others) must be understood as a part of the authoritarian package. Altemeyer also realized, as had earlier researchers, that authoritarianism was less about political ideology, per se, than it was about narcissism, fear, otherness, and violence; as the examples of Stalin and Hitler demonstrate, authoritarians can flourish on the extremes of both the left and the right. In their *Authoritarian Nightmare: Trump and His Followers*, Altemeyer joins political commentator John W. Dean to discern and analyze two major types of authoritarians—social dominators and authoritarian followers. Before the 2016 election, Altemeyer predicted that Trump could win because studies he had been doing since 1981 had led him to conclude that "there were so many people so genuinely submissive to established authority [among these two groups] that they constitute a real threat to freedom" in the US. (Dean and Altemeyer 2020: 104).

Altemeyer's first authoritarian group, the "social dominators," believes that some groups deserve to be "more prestigious and powerful than others" (2020: 108). This psychological-political type, many of them mean-spirited males, desire personal power and will practice cheating, manipulation, lying, intimidation, and bullying to achieve it. They saw one of their own in Trump and enjoyed his norm-busting, abusive, and disruptive campaign for the presidency. Like Trump, they much prefer "winners" to "losers," but will line up behind a strong man if they can't be winners themselves in the expectation of basking in some of his success. Inherently racist and domineering, they were emboldened by Trump's viciousness; many had been "waiting for a leader who would let them unleash a hatred that has long roiled within them," notes Altemeyer (122).

His second type, "authoritarian followers," typically embody "high degrees of submission to the perceived established legitimate authorities in society," "high levels of aggression in the name of their authorities," and "a high level of conventionalism," which leads them to follow authoritarian norms (Dean and Altmeyer 2020: 124). Many of these right-wing authoritarian followers practice evangelical Christianity or conservative versions of Catholicism. To the statement, "Our country desperately needs a mighty leader who will do what has to be done to destroy the radical new ways and sinfulness that are ruining us," most authoritarian followers respond with "very strong agreement" (2020: 126). In assessing the thought patterns of this psychological group, Altemeyer found them "highly ethnocentric," "dogmatic," "highly prejudiced," and prone to "conflicting and contradictory ideas" (136–149). Both social dominators and authoritarian followers dismiss the importance of civil liberties for all, display aggression against those with whom they disagree, demand constant reinforcement from followers and others for their beliefs, and are uniquely vulnerable to manipulation from leaders they admire. These qualities, shared by most of Trump's political base—roughly 40% of the voting population—made it likely that they would continue to support Trump regardless of his policies and refuse to abandon him in the 2020 election.

Trump's affinity for undemocratic practices during the 2016 election played out in his administration. To return to Levitsky and Ziblatt's normative understanding of the death of democracies, the authors concluded that "President Trump exhibited clear authoritarian instincts during his first year in office" (Levitsky and Ziblatt 2018: 177). The authors list three significant norm-breaking strategies that authoritarians around the world have deployed to consolidate power—"capturing the referees, sidelining the key players, and rewriting the rules to tilt the playing field against their opponents" (2018: 177)—and find that President Trump used all three of them in his first twelve months in office.

Included among the "referees" that Trump attempted to capture were officials in law enforcement, the intelligence agencies, bureaucratic whistleblowers for ethics, and the courts. Soon after assuming office, President Trump pressured Director of the FBI James Comey to pledge his loyalty to him and drop the investigation into the actions of Michael Flynn, his initial Director of National Security, who was later found to have violated ethics codes and collaborated with the Russians during the election. When Comey refused to comply, Trump found an excuse to fire him. In addition, the President used Twitter to attack the Judge who initially voided his administration's travel ban for Muslims and other proscribed groups of travelers. Other judges who ruled against Trump's policies suffered similar Twitter tirades.

Trump also "sidelined key players" and attempted to "rewrite the rules" of fair play during his first year. His attack on mainstream news critics is a good example of this sidelining strategy. Trump's "repeated accusations that such outlets as the *New York Times* and CNN were dispensing 'fake news' and conspiring against him look familiar to any student of authoritarianism," say Levitsky and Ziblatt (2018: 181). Regarding Trump's rewriting of the "rules of the game," his response to the White-supremacist march in Charlottesville, VA in August stating that there were "good people" on both sides is a clear example of what Levitsky and Ziblatt would call (to continue their sports metaphor) moving the moral goal posts by equating law-abiding left-wing protesters with murderous neo-Nazis. The authors add that "perhaps the most antidemocratic initiative yet undertaken by the Trump administration" was his attack on the right of racialized and ethnic populations to vote by creating the Presidential Advisory Commission on Election Integrity (183).

Because they published *How Democracies Die* early in 2018, the authors did not deliver a report card on Trump's push toward an authoritarian presidency in that year or the next. Nonetheless, we can take Levitsky and Ziblatt's three major strategies that authoritarians use to undermine democratic norms and find numerous instances of their deployment in 2018 and 2019. Trump's many attacks against the rule of law, for example, continued his attempts to "capture the referees." For twenty-two months, he tried to close down, undercut, and/or silence Special Counsel Robert Mueller's investigation into Russian interference in the 2016 election. Mueller's investigative team indicted thirty-four people, roughly a third of them from Russia, and handed off several related lawsuits to other units in the Justice Department. Although Mueller did not follow the money to uncover possible financial crimes and compromises, his final Report, completed in March of 2019, listed ten instances of possible obstruction of justice by candidate-then-President Trump from the campaign through the first year and a half of his presidency. These included bribery, witness tampering, illegal offers of a pardon, and the attempt by Trump to get White House Counsel, Don McGahn, to fire Mueller.

In addition to sidelining the Special Council, the Trump administration also disempowered Congress, Republicans included. In December of 2018, after both houses of Congress patched together a bipartisan spending deal on the budget, the President torpedoed the bill because, in his mind, it did not contain enough money to build a border wall between Mexico and the US. Soon before the publication of their book, Levitsky and Ziblatt wrote an essay for *The New York Times* excoriating Trump for this breach of democratic norms. "These developments should set off alarm bells," they conclude. "Our president is behaving like an autocrat. His willingness to fabricate a national crisis and subvert constitutional checks and balances to avoid legislative defeat places him closer to Ferdinand Marcos than to Ronald Reagan" (Levitsky and Ziblatt 2018: December 23). When Trump failed to get money for his wall through the shutdown, he shifted congressionally allocated funding out of the budgets for Defense, Homeland Security, and FEMA and into a special fund for his pet project.

Compounding Trump's attempts to capture the referees of democratic fair play was his continual lying. Near the end 2019, fact-checkers for the *Washington Post* calculated that Trump had passed the 15,000 mark in terms of telling outright lies to the public. This break with democratic norms reached over 30,000 lies by the day he left office in 2021. Further, as rhetorician Jennifer Mercieca notes in her *Demagogue for President*, Trump used the rhetoric of authoritarian control to shape many of his lies. During the campaign and throughout his presidency, he pandered to an imagined popular will he had created by using a "many-people-are-saying" strategy to frame his lies as an appeal to the wisdom of the crowd. He also issued threats of intimidation and brute force; encouraging crowds at his rallies to chant "lock her up" against Hillary Clinton was perhaps the most obvious example. Finally, Trump enjoyed leading his followers down tortuous rabbit holes of endless conspiracies, all the while professing amazement about "what's going on here?" When a commonly agreed upon body of facts can no longer be taken for granted in public debate, citizens become cynics, attempts at compromise collapse into rancor, and people either turn away from public participation or embrace the authoritarian con man as their savior. The Big Lie technique that Hitler used in Germany in the 1930s—repeat the lie long and insistently enough and your followers will believe it—was alive and well in the US under Trumpism.

Although several news outlets frequently commented on Trump's mendacity, Fox News, the most-watched cable news network during Trump's presidency, was not among them. In a well-researched essay for

The New Yorker published before the release of the Muller Report, investigative journalist Jane Mayer blasted Rupert Murdoch, the owner of Fox, and others at the network for transforming their news operation into a propaganda factory for Trump. The journalist cites numerous instances of Fox reporting that backed up Trump's lies with the network's lies to their audience. Mayer quotes the cofounder of a moderate think-tank in Washington DC for her conclusion: "In a hypothetical world without Fox News, if President Trump were to be hit hard by the Muller report, it would be the end of him. But with Fox News covering his back with the Republican base, he has a fighting chance, because he has something no other President in American history has ever had at his disposal – a servile propaganda operation" (Mayer 2019: 53). In effect, Fox was to Trump's popularity what Nazi radio and Riefenstahl's *Triumph of the Will* were to Hitler.

In addition to conspiring with the most popular news operation in the nation to uphold his lies and inflate his popularity, Trump "sidelined" many governmental officials, the second of Levitsky and Ziblatt's antidemocratic norm-breaking strategies. These included most of the cabinet chiefs, top bureaucrats, and scientists in government who might oppose him on policy matters or leak information damaging to his administration. True to his authoritarian suspicions (and to Republican fears of Big Government), Trump treated Washington bureaucrats as vanquished enemies who must be prevented from regaining power and disabling his presidency. After three years, most of his initial appointments to cabinet-level positions had left and their positions were filled by toadies with little authority of their own to run their departments or report to Congress. They could also be replaced by the Tweeter-in-Chief if they angered the President. According to reporter Evan Osnos, numerous vacancies, especially in the State Department and the EPA, were left unfilled and some agencies, including Housing and Urban Development and the Department of the Interior, avoided creating public records so that investigators from Congress would have nothing to examine when they questioned their procedures and results (Osnos 2018: 56–65).

Levitsky and Ziblatt's final criterion for tracking the rise of authoritarianism in the US is the president's ability to "tilt the political playing field" against his opponents. Intimidating potential Democratic groups from voting was one such strategy. During the Jim Crow era in the South, a lynching near election time sent a clear message to African Americans. White-supremacist hate groups continued to send this message to minorities and they were joined by a variety of other right-wing extremists targeting US Jews, Latinos, and Muslims as well as Mexicans and other groups from Latin America whom they suspected of illegally crossing the southern border of the nation to vote. Authoritarian hate groups surged during Trump's campaign and after his election; the Southern Poverty Law Center (SPLC), which monitors the activities of such groups, tracked a rise in the number of hate groups from 784 in 2014 to 1020 in 2018. According to SPLC statistics, these alt-right domestic terrorists, several of whom used the same kind of racist language as the President to describe their targets, killed 57 people and injured 80 others in 14 deadly attacks during the first two years of Trump's presidency.

President Trump doubled down on his racist strategy to denigrate non-White Americans and attack immigrants from Latin America before the 2018 midterm elections. In numerous tweets, rallies, and Fox News reports, he stoked fears about a lawless "caravan" of migrants from Central America making their way across Mexico to crash through the southern border of the US and invade the nation. Evidently, Trump and Fox's attempts to ramp up xenophobic hysteria before the election failed to convince many undecided voters, however. Although Republicans expanded their small majority by two votes in the Senate, the Democrats gained 41 seats in the House of Representatives, won seven state governorships and at least 350 state legislative seats, and took control of six state legislative chambers. This "blue wave" election, which marked the highest voter turnout for midterms since 1914, not only left the Dems in control of the House but also passed laws raising the minimum wage in several states.

The defeat apparently convinced Trump and some of his associates in the White House that they must tilt the playing field at an even sharper angle than they did in 2016 to win the presidential election in 2020. Although the Mueller Report presented substantial evidence that Trump clearly benefitted from Russian interference, the President neither admitted Russian interference, nor disavowed seeking electoral help from other nations in the future. The reason is not difficult to fathom. As Jamelle Bouie noted in an op-ed piece for *The New York Times*, the Republicans understand that, given the demographic trends in the nation, their party may be doomed to minority status in future presidential elections. Consequently, they have decided that "the only way to preserve right-wing conservatism in American government is to rig the system against this new electorate" (Bouie 2019: April 28). This has led many

Republicans to reject all efforts aimed at electoral expansion, such as early voting, mail-in ballots, and automatic registration. State house Republicans have also supported gutting the provisions of the Voting Rights Act of 1965 to enact voter-ID laws in order to restrict access to the ballot from non-White minorities and the poor. Bouie concludes that the new Republican strategy for electoral victories in the future is simple: "If you can't win playing by the rules, you change them" (2019: April 28).

The Democrats caught Trump trying to change the rules in violation of his Oath of Office and the rule of law in mid-September of 2019, when a whistleblower's report surfaced that credibly implicated the President in a conspiracy of extortion against the President of Ukraine. Together with his personal lawyer Rudi Giuliani and some American diplomats, the Secretary of State, and even the Vice President, Trump pressured President Zelensky to fabricate a legal case against former Vice President Joe Biden, then the likely candidate to run against Trump in 2020. The partial transcript of a July phone conversation between Trump and Zelensky released by the White House confirmed the whistleblower's accusation that Trump was withholding promised military assistance to Ukraine—then in the midst of fighting a five-year war against Russia for its occupation of Ukrainian territory—in order to leverage dirt on Biden. As several Democrats and commentators noted, Trump's strategy was a classic Mafia shakedown. Subsequent inquiries confirmed the essential truth of the whistleblower's accusation and the Democrats in the House moved quickly to begin impeachment proceedings against the President.

Despite Trump's order to subordinates in his administration to refuse subpoenas from the House Judiciary Committee, seventeen witnesses testified, risking their careers to defend the Constitution. In the end, the Committee sent two articles of impeachment against President Trump to the heavily Democratic House, which voted overwhelmingly to approve them. The first article castigated the President for using public money to strong-arm a vulnerable ally and the second rejected the legality of his categorical defiance of Congress in refusing to comply with the constitutionality of their investigation. Minority Republicans in the House and those controlling the Senate echoed Trump's objections of "witch hunt," "conspiracy," and "sham" and many refused to admit that the President had done anything wrong at all. Not surprisingly, voters were nearly as polarized as the two parties. The trial to remove President Trump from office began in the Senate early in 2020. As expected, the Republican-controlled Senate acquitted Trump of both charges and, to judge from his polling, Trump's popularity among his pro-authoritarian base barely budged.

Throughout the first three years of Trump's presidency, opponents in the liberal press often asked when individual Republicans would grow a backbone and wake up to Trump's authoritarian power grabs. The liberal media's rationales for the apparent indifference of most Republicans to Trump's behavior ranged from simple cowardice to political savvy, but few journalists accused them of sharing many of the same politically charged personality traits as the President. Dean and Altemeyer's *Authoritarian Nightmare*, however, challenges this conclusion. Altemeyer and his colleagues had already conducted eight different surveys of over twelve hundred lawmakers in state legislatures between 1990 and 1993 to test for indicators of authoritarianism among them. He found that nearly all Republican state lawmakers (except in Connecticut) demonstrated moderate or extreme authoritarian traits, whereas most Democrats were anti- or only moderately authoritarian in their attitudes. Altemeyer teamed up with the acclaimed Monmouth University Polling Institute in October and November of 2018 to ask nearly a thousand registered voters their political preferences, question them about child-rearing and religious fundamentalism, and administer five personality tests. The results supported the conclusion that Trump voters in 2018 were "highly authoritarian;" his White supporters were also "highly prejudiced" and most of this prejudice can be "explained by authoritarianism" (Dean and Altemeyer 2020: 216–219). Republican enablers of Trumpism were not simply spineless fools or wily politicos; most of them, like their hero, were and are deeply authoritarian.

8.2 Polarized Films, 2000–2012

The social scientists I have relied upon in this chapter offer empirically solid psychological, social, economic, and political explanations for the rise of Trumpism, but they slight the causal efficacy of the political stories that Americans have been telling each other through film and other media for the last twenty years. In preparation for this discussion of polarized US films, I will return to Tuschman's coevolutionary perspective on the political left and right. Recall that Tuschman bases his typology primarily on two of the "Big Five" personality dimensions that have been well-tested and widely accepted as indicators of political orientation—openness and conscientiousness. Openness aligns closely with left-wing values and voting and conscientiousness with the right wing. Taking empirical data on other personality factors (including indicators for authoritarianism) into consideration, Tuschman derives three general factors for voters that play out universally across the world on the left-right spectrum: tribalism, tolerance of inequality, and competing perceptions of human nature. Left-wing voters are generally less ethnocentric, less religious, less intolerant of non-reproductive sexuality (all indicators for tribalism), less tolerant of inequality and hierarchy, and see others as less competitive than centrist voters. In contrast, right-wing voters are generally less tolerant of other groups, less secular, less accepting of non-reproductive sexuality, less intolerant of inequality and hierarchy, and see others as less cooperative than voters in the center.

With these distinctions in mind, we can trace four rough categories of Hollywood dramatic films over the past twenty years or so. The first three are simply right wing, left wing, and centrist films, with non-political films—movies that either have little to do with tribalism, inequality, and extremist perceptions of human nature or invite contradictory and/or confusing emotional responses to these politically charged orientations—as the fourth (and largest) category. My goal for this section is to summarize major trends in popular right- and left-wing dysphoric films in the US from around 2000 to 2018, using Tuschman's indicators. As will be evident, these dysphoric political films have been both a product and a partial cause of the increasing polarization of the US polity for the last twenty years.

I will look to Peter Biskind's entertaining and generally perceptive book, *The Sky is Falling: How Vampires, Zombies, Androids, and Superheroes Made America Great for Extremism*, to guide us into the contemporary world of politics and film viewing. For the most part, Biskind discusses popular mainstream movies, produced by major studios with well-known directors, and featuring recognized stars. His introduction gives an accurate and credible account of the political culture of American films in the pre-2000 era before citing some of the same realities noted in past chapters that have moved over two hundred popular films and television series toward the political extremes since the turn of the century.

As Biskind notes, "Extremism has defined today's entertainment culture to such an extent that no one gives it a second thought – certainly not the studios – save for exceptions like *New Yorker* reviewer Anthony Lane" (Biskind 2018: 12). Biskind quotes the opening of Lane's review of Tarantino's *The Hateful Eight* and Iñárritu's *The Revenant*, for example, both of which opened on Christmas Day of 2015 and clearly embody what Tuschman understands as a violent and competitive view of human nature: "Five hours and thirty-eight minutes of malice and mistrust, in which the characters – mostly men – are trapped in extreme weather conditions and settle their differences with extreme violence. So much for peace and good will" (quoted in 2018: 12). Although he primarily groups films according to their ideological content rather than Tuschman's psychology and Whitehouse's concerns with emotional response, his loose categories of left and right often finds informal ways of taking my orientations into account. Tellingly, however, as will be evident, Biskind ignores a major recent shift in the politics of film viewing that derives from the depiction of African American slavery.

Partly because Republicans were pushing their party toward extreme right-wing positions before the Democrats did so on the left, it is not surprising to find that extremist films on the right flourished among viewers earlier than their polar opposites. As noted, Republican authoritarianism in the US appeals to two major blocks of voters, authoritarian followers, and social dominators. I will briefly discuss a group of films that have been popular with evangelical Christians, who are predominately authoritarian followers, before turning to the more mainstream films that have likely appealed to social dominators as well as other demographics. Beginning with many of the films of D. W. Griffith, there has always been a US market for movies with Christian stories and themes. The huge box office success of Mel Gibson's *The Passion of the Christ* (2004), which combined evangelical Protestant with conservative Catholic themes and spectacle, awakened many filmmakers to the growing audience of conservative Christians. Several TV series and a string of low-budget Biblical pics followed, including *God's Not Dead* (2014), which, Biskind reports, took in $61 million in the US on a $2 million budget.

The *Left Behind* TV and film productions, based on sixteen fundamentalist thrillers about the Rapture, have been the longest running of the evangelical series. In the first film, titled simply *Left Behind* (2000), spectators know the End of Days has arrived because God saves Jerusalem from destruction when planes attacking Israel unaccountably fall out of the sky, even before the initial film credits have run their course. Soon after that, God's Chosen vanish from the Earth, raptured out of their beds, from behind their desks, and up from their seats in airplanes. They leave behind their clothes, wedding rings, false teeth, and their surprised relatives who (apparently) weren't good enough to get into heaven on the first try. Also left behind to rule the Earth is the Anti-Christ in the figure of Carpathia, who soon declares himself the new Secretary-General of the United Nations and vows to cause starvation and plague for seven years, as the Almighty promised in the Book of Revelations. Although the disappearance of the raptured is understood primarily as a blessed euphoric event, the film turns sharply dysphoric when Carpathia begins killing people to ensure his rule, confirming the religious-Right's suspicion of UN authority and other international restraints on US power. With the Almighty driving the plot, however, the fundamentalists know that God is in control, even if this means seven years in hell for the left-behinds.

The main outline of the *Left Behind* plot changed little after 2000. A 2014 version of the story included this beginning dialogue to flag the coming apocalypse: "Matthew 24, Verse 7 says there's going to be famines, pestilence, and earthquakes in diverse places." "Then why doesn't He do something? He's God right?" "It's a fallen world. God created it perfect and we destroyed it. With the first sin" (quoted in Biskind 2018: 50). Because the old truths are the only truths, what's to change? The Bible already spelled out the major plot point of the Rapture story and Hollywood is happy to echo its variations: God's fearful destruction is coming, but some of us will be saved. The *Left Behind* films repeat the patriarchal logic of manuscript culture uttered by C. B. DeMille's Pharaoh in *The Ten Commandments* in 1956, "So let it be written, so let it be done."

Biskind traces many of the early themes popular with what Dean and Altemeyer call social dominators to *Independence Day* (1996), a mostly dysphoric film in which the President of the US, faced with the invasion of aliens in a mother ship 340 miles in diameter, believably states that we are no longer battling against "tyranny, oppression, or persecution," we are fighting against "annihilation" (quoted in Biskind 2018: 24). As this quotation suggests, the Cold War tethered much of Hollywood's imagination to climactic battles with the Russians; the cadets in *Red Dawn* (1984), for example, must repel a Communist military invasion. Nonetheless, the Manichaean morality and macho muscle of many Cold War films, especially those like *Patton* that harken back to the last "good war" the nation fought, echo throughout *Independence Day*. Regarding the intergalactic villains of the film, many on the authoritarian Right now see the displacement of White privilege from normative power in the US as a form of "annihilation" and view their Black and brown opponents as aliens. Indeed, with their armor plating removed, the aliens of the film are similar to the zombies in *Night of the Living Dead*—globs of slime with eyes, mouths, and tentacles that look hungry and infectious.

Eventually opposing them, after crushing defeats by the aliens, are a rag-tag air force of militiamen heroes, one of whom, a former Vietnam vet pilot, altruistically sacrifices his life to save US-led humanity. With more vengeful violence than even Coppola's Corleone's could mount against their Mafiosi enemies at the conclusion of *The Godfather*, *Independence Day* is also deeply intolerant of non-reproductive sex. Ridiculous gay scientists, almost as freakish as the aliens themselves, are contrasted with heightened heteronormative males who share cigars after combat, among them the fighter-pilot President of the US. Even in the midst of its apocalyptic-level chaos, the movie preserves the hierarchy of military command from the President down to captains and corporals.

Director Brad Bird started a trend in libertarian films with *The Incredibles* in 2004, an animated, family-friendly introduction to Ayn Rand's world of superheroes whose liberty is compromised by foolish governmental authorities. Upset with the collateral damage caused by the good deeds of Mr. Incredible, his wife Elastigirl, and their superchildren, the authorities ban the family and force them to join the Superhero Relocation Plan. The real villain in this satire is the US government, which enforces mediocrity through democracy. Bird moved from the euphoria of *The Incredibles* to the grim dystopia of *Tomorrowland* (2015), which depicts a future of famine, wars, and overpopulation, all because America has been deprived of the genius of its greatest artists, scientists, and inventors. As Biskind remarks, Tomorrowland, the institution at the center of the film charged with designing a better tomorrow, "is a sort of futuristic Heritage Foundation for the best and the brightest" (Biskind 2018: 54). Also fixated on a libertarian (and imperialistic) future, *Interstellar* (2014), directed by Christopher Nolan, challenges humanity to explore and exploit distant galaxies.

By this point in his career, Nolan was better known for his *Dark Knight* trilogy of Batman films, a more direct descendent of *Independence Day* than the libertarian films above. Gone from the trilogy are the jokey plots, silly music, and enjoyable villains (Danny DeVito's Penguin, among them) of earlier Batman movies. Although more euphoric than the last two films in the series, *Batman Begins* (2005) orients the series toward urban crime, vengeance, vigilante justice, and neoliberal economics. As a child, Bruce Wayne watches his parents murdered as they exit from a movie theater in Gotham and he vows to avenge their deaths. Years later, traveling the world as a rich young man, Wayne lands in a Bhutan jail and an oriental sage with Bruce-Lee fighting skills rescues him from death at the hands of local thugs by training him in self-defense. After the two of them break out of the prison, Wayne discovers that his mentor is the leader of the League of Shadows, a band of assassins and vigilantes, who tells him, "You have the strength of a man denied revenge" (Nolan et al. 2005). The League orders Wayne to return to his hometown to help the League destroy it because "cities like Gotham are in their death throes – chaotic, grotesque, beyond saving" (Nolan et al. 2005).

Sure enough, Gotham is a cesspool of criminality and urban decay, run by a mob boss in league with a corrupt psychiatrist who controls the city's asylum. But Bruce encounters several problems upon his return to Gotham. Just as he is about to pull the trigger on his parents' killer, another man assassinates him first. Then his childhood sweetheart, still a possible love interest, explains to him that justice is better than revenge. And finally the Mayor of the city seeks help from Bruce, now the head of Wayne Enterprises and the richest man in Gotham, in reforming the urban swamp. What is a poor proto-superhero to do? The first film of the trilogy tries the path of neoliberal philanthropy and civic responsibility, but finds that it pinches Batman's potential as a vigilante do-gooder. Without ever explaining how Wayne Enterprises could be so wealthy and not have helped to cause the degradation of Gotham, Bruce Wayne chooses a double life: he will give up his playboy ways to help the city financially and secretly turn himself into Batman to rescue the good people in the movie and turn the bad people over to city justice. And he does.

By the end of *Dark Knight Rises* (2012), the final film of the three, however, these good intentions are in tatters. The second film, *The Dark Knight* (2008), is primarily built around the villainous plans of the Joker, an amoral narcissist jealous of Batman's fame who seeks attention through spectacular murders. As played by Heath Ledger, the psychopath commands almost as much screen time as Christian Bale's Bruce Wayne/Batman. He certainly excited more fear and, judging by the reviews, probably admiration. (In addition, Ledger's death at the end of the filming substantially bumped up the popularity of *Dark Knight* after its release.) The second film also features the gradual disintegration of Harvey Dent's civic heroism. A classic good cop who cracks under the pressure of Gotham's corruption and the Joker's shameless antics, Dent goes on a killing spree and must be saved by Bruce/Batman to protect the reputation of the Gotham police.

As Biskind notes, "vigilantism is a growth industry" over the course of the trilogy (Biskind 2018: 159). In *Dark Knight Rises* (2012), Batman must save Gotham from destruction by the League by freeing the city police trapped in the sewers, returning the poor and the lunatics of the city to safety, overcoming the wiles of a femme fatale working for the League, and neutralizing the explosion of a neutron bomb. At the end of *Rises*, a promising young detective is offered a promotion in Gotham's police department, but turns it down because he has learned that breaking the law to kill criminals is more effective than enforcing it. Says Commissioner Gordon, "There's a point when the structures fail you and the rules aren't weapons anymore; they're shackles letting the bad guy get ahead" (Nolan et al. 2012). The primary writer for the series, Frank Miller, modeled the role of Batman on Clint Eastwood's Dirty Harry, featured in five films in the 1970s and 1980s. Fascinated by the explosive violence of Eastwood's persona, Miller approved of Dirty Harry's willingness to administer murderous vigilante justice to bad guys and his script for *Rises* follows that code. The moral of the trilogy is clear: If human beings living in corrupt circumstances are understood as naturally competitive and violent, it is foolish to play by the rules.

As right-wing films veered toward hypermasculinity, vigilante authoritarianism, and the perception that human nature was invariably competitive and violent, the left-wing eventually found its filmic voice by moving forthrightly in the opposite direction. Biskind recognizes James Cameron's 2009 blockbuster, *Avatar*, as the left-wing's response to the right's *Independence Day*. The film embraces pliable sexuality, transspecies flexibility, and a belief in human nature as normally compassionate and peaceful. Along the way, *Avatar* also rejects a world dominated by machines, especially the machines of war, for an affirmation of organic life and ecological balance. In an interview with a *New York Times* critic, Cameron acknowledged that he set out to fulfill a progressive political agenda. Citing the importance of documentaries like Al Gore's *An Inconvenient Truth*, Cameron admitted he wanted the viewers of *Avatar* "to be warriors for the earth" (quoted in Biskind 2018: 98).

Avatar also drew inspiration from Kubrick's *2001: A Space Odyssey* (1968) and several Center-Left films about the dominance of technology, friendly alien otherness, and capitalistic greed from the 1970s and 1980s. In *2001*, a homicidal computer, the HAL 9000, takes over a manned spaceship and kills most of its human operators. *Avatar*, however, also nods to Spielberg's kid-friendly sci-fi aliens in *Close Encounters of the Third Kind* (1977) and *E.T. the Extraterrestrial* (1982), as well as such anti-capitalist, dystopic thrillers as *Blade Runner* (1982) and *Brazil* (1985). Cameron had already directed a sci-fi thriller, *Aliens* (1986), a sequel to *Alien*, produced seven years earlier, that continued the story of a female commander's attempt to exterminate a colony of ravenous monsters who incubated in human bodies and threatened to take over the Earth. In the shift from *Aliens* to *Avatar*, Cameron kept his critique of technological progress and capitalist oppression, but jettisoned the fear of alien otherness.

Cameron aims *Avatar* at extractive capitalism, American imperialism, and neoliberal globalization. Set 150 years in the future, *Avatar* imagines the voyage of a freight train-like spaceship that sets out from a plundered Earth to capture the resources of Pandora, a ripe and inviting utopia inhabited by the Na'vi, who initially live peaceable and cooperative lives. The operation, run by capitalist boss Parker Selfridge, is prepared to use "avatars"—Na'vi bodies scientifically "driven" by individual US minds— to convince their compatriots to give up their resources to the invaders. As the capitalist-imperialist explains to the scientists running the avatars, "Just find me a carrot that will get them to move, otherwise it's going to have to be all stick" (Cameron 2009), by which Selfridge means the well-armed Marines on board the US government spaceship sent to enforce its extractive mission. Jake Sully, a former Marine who needs a wheelchair due to a battle injury when he's not running one of the avatars, however, refuses to go along with the plan. He falls in love with a Na'vi girl, who teaches him to love the exotic natural beauty of Pandora, which leads the leader of the US Marines to bully Sully for "betray[ing] your own race" (Cameron 2009). After fighting breaks out between the Na'vi and the Marines, Sully, already part-Na'vi from his avatar experiences, must transform himself into a full Na'vi in order to breathe the atmosphere of Pandora and resist the Marines' imperialist invasion. So he takes the plunge and Na'vi magic transforms his species-based DNA from one of "us" to a member of "them." Take that, *Independence Day*!

When *Avatar* was released in 2009, left-wing and mainstream reviewers were ecstatic in their praise; the film generated nine Oscar nominations and box office receipts topped Cameron's previous success with *Titanic*, generating 2.8 billion by 2018. The movie generated several others with much the same orientation, among them the four films of the *Hunger Games* series (2012–2015). These flicks paraded a dystopian version of the original *Star Wars* trilogy (1977–1983), minus George Lucas's mythic heroic past of Jedi knights and the updraft of optimism that *Star Wars* received from the success of the US space program. Both series feature a teen protagonist from the backwoods whose quest for justice takes him or her (Katniss Everdeen in *Hunger Games*) from a hum-drum provincial existence into new dangers, romantic conflicts, and an eventual confrontation with oppressive imperial power. Instead of help from Obi-Wan Kenobi and The Force, however, Katniss has only her family and a village drunk, Haymitch (engagingly played by Woody Harrelson), to advise her initially in the backwater of District 12, a rough parallel to Appalachia in the 1950s. The series initially pits selected teenagers from various tribute states against one another in a deadly survival of the fittest reality show sponsored by an authoritarian state run for the amusement and benefit of US oligarchs in a future dystopia called Panem. Katniss, an excellent archer, chooses to enter the games in the place of her younger sister, who was originally chosen to compete. With her male partner from District 12, she struggles against the game's rules as well as her opponents in the arena and wins the deadly contest for both of them. As in *Avatar*, the target of the series is an oppressive oligarchy and its implicit demand for competition over the teens' natural preference for peace and cooperation.

To make sure the audience links teenage rebellion against the corrupt rich to a possible future revolution, the second of the four-part series, *Hunger Games: Catching Fire* (2013), arranges early on for President Snow, the dictator of Panem (played with creepy unctuousness by Donald Sutherland), to ask Katniss, "What is to prevent an uprising that can lead to revolution?" (quoted in Biskind 2018: 144). Recognizing that Katniss could become a symbol of revolt, Snow changes the rules to pit previous winners of the Games against each other, in the hope of eliminating her threat to his power. Elevating Katniss to Liberty Leading the People betrays the origin of the series in three adolescent novels, but the actor playing the role, Jennifer Lawrence, does manage to depict a gravitas beyond her years for this impossible role. By the end of *Catching Fire*, all of the Districts in President Snow's fascist state are in rebellion against the Capitol. In the third film, *Mockingjay, Part I*, the least successful of the four, the rebels mount an offensive, but Snow and his army prevail. Critics disappointed in Part I of *Mockingjay*, however, could be sure that Hollywood would provide an expensive ending to the series in Part II because the producers had hired a bevy of well-known stars for both parts of the film. These included Philip Seymour Hoffman, Stanley Tucci, and Julianne Moore, who plays the rebel President Alma Coin.

As promised, *Mockingjay, Part II* delivers a dark and exciting story. After avoiding several traps set for them in the outer rings of the Capitol, Katniss and her multicultural mates crawl through the city's sewers—think a combination of the WW II sewers of *Kanal* in Warsaw and those of *The Third Man* in Vienna—and what's left of her platoon emerge near the President's Winter Palace at the center of the city to join the trudge of citizens toward Snow's offer of a safe haven for Capitol families from the invading rebels. A loudspeaker urges the families to pass their children to the gates of the palace, where they will be allowed to enter, and as this is occurring small parachutes descend from planes, hover over the children's heads, and begin exploding, killing everyone below them. Katniss barely escapes as the remaining citizens rush the palace in anger to arrest Snow and declare victory for the people. In a later private conversation with Snow, however, Katniss hears that the rebel President Coin actually ordered the death of the children to provoke Snow's overthrow and seize power herself. Coin unknowingly confirms Snow's accusation when she declares herself to be the "Interim President" of the nation and asks Katniss to publicly execute Snow for his crimes against humanity. At the climax of the film, Katniss takes aim at Snow, but shoots her arrow a little higher and kills President Coin standing above the villain. With the rebels finally victorious, peace and love abound; Katniss and her former boyfriend from District 12, now presumably married, relax near a river with their two children as romantic music and visions of Nature move the audience toward The End.

Avatar and the four *Hunger Games* films successfully countered the authoritarian assumptions underlying many of the most popular right-wing films from *Independence Day* through *The Dark Knight Rises*. They did so both by challenging them directly and by substituting other notions about humanity that were beginning to remake popular ideas of human nature. The authoritarian assumptions embedded in *Independence Day* and the *Dark Knight* series included the notion that natural men are aggressive and domineering, that ordinary people could be easily manipulated to torture their fellow humans, and that governmental power could do little to inhibit the inherent selfishness of humankind. These assumptions, of course, also justified the neoliberal order, with its vast disparities of wealth, racist caste system, and limitations on political democracy.

During these same years, however, many social scientists, animated by new research in several fields, including coevolution, were dismantling many of these neoliberal premises. In his *Humankind: A Hopeful History*, Rutger Bregman outlines several of these significant changes in what he, along with many others, simply terms "human nature" (Bregman 2019: 2). Calling his point of view "a new realism," Bregman announces the book's thesis early in his introduction: "This book is about a radical idea…. An idea so intrinsic to human nature that it goes unnoticed and gets overlooked…. So what is this radical idea? That most people, deep down, are pretty decent" (2019: 2). And Bregman returns much of his discussion to Pleistocene realities to prove it.

Bregman's chapters in his Part 1, entitled "The State of Nature," elevate Rousseau's vision of equality over Hobbes's argument for a Leviathan, reveal research about the purported aggressiveness of soldiers, and debunk the notion that civilization is simply a thin veneer masking primeval natural aggression. Reliable reports and interviews during and shortly after WW II revealed that most soldiers in the midst of combat do not fire their weapons because it is difficult to get most people, even in life-threatening circumstances, to kill other human beings. Bregman also cites extensive anthropological research into the origins of war, which substantiate the conclusions of coevolutionary scientists previously cited that warfare, mostly absent in the Pleistocene, was an invention of agricultural civilizations. In Part II's "After Auschwitz," Bregman looks closely at Stanley Milgram's experiments at Yale University in the 1960s. Milgram, who presented his research as an explanation for how human beings could have caused the Holocaust, concluded that our species has a fatal flaw—we act like obedient puppies, summarizes Bregman, when ordered by legitimate authorities to hurt other people. Bregman shows, however, that Milgram exaggerated his findings. Later researchers trying to replicate Milgram's experiments found that those asked by the "scientists" in white coats to torture others would only do so willingly if they believed they were doing good. "In fact, people go to great lengths, will suffer great distress, to be good," noted a psychologist who attempted to repeat the circumstances of Milgram's Holocaust-like simulation (2019: 169). A chapter entitled "What the Enlightenment Got Wrong" in Part III critiques Adam Smith, David Hume, James Madison, and other major moralists of the period for advancing the idea that economic greed and political ambition can be made to advance civilized polities through markets and political institutions that turn these apparent evils into public goods. Realizing that many of the major institutions of western capitalism and governance continue to rely on these enlightenment assumptions, Bregman asks whether we can invent new institutions "that operate on a wholly different view of human nature?" (2019: 250).

Although *Avatar* and the *Hunger Games* series are too dysphoric to attempt the building of new institutions, both undermine authoritarian assumptions and suggest a different natural morality upon which more just institutions might flourish. To borrow Bregman's characterization of most of our species, the protagonists of these two films, Jake Sully and Katniss Everdeen, are, "deep down, pretty decent" people. Both repeatedly choose cooperation and altruism over domination. On the scale of hierarchy vs. egalitarianism, Katniss, raised in a poor Appalachian family, has a keen sense of fairness that pulls her toward sympathy with underdogs who have less power and status than others. Jake, a soldier in a wheelchair, understands what it means to be an outsider and easily practices xenophilia rather than tribalism in his relations with alien others. Having suffered unjust oppression from the powerful, both know that power corrupts and that powerful people tend to be more disconnected, cynical, and lacking in empathy than others. In his *Humankind*, Bregman lists the primary traits that, according to anthropologist Chris Boehm, the leaders of hunter-gatherer groups needed to flourish. For most of human life on Earth, he says, such leaders needed to be "Generous, Brave, Wise, Charismatic, Fair, Impartial, Reliable, Tactful, Strong, [and] Humble" (Bregman 2019: 231). Bregman demonstrates, however, that since the transition to agriculture, especially after the invention of writing facilitated the flourishing of propaganda, most leaders have needed to specialize in one trait above all others: shamelessness. He adds, "Such individuals wouldn't last long in nomadic tribes. They'd be cast out of the group and left to die alone. But in our modern sprawling organizations, sociopaths actually seem to be a few steps ahead on the career ladder" (2019: 239).

Another way to understand these left-wing films is to place them in the context of Obama's first term as President, when the qualities of Pleistocene prestige and leadership actually seemed to be embodied in a President who was trying to move the US toward a progressive future. In the wake of Obama's early successes, his empathy in the face of numerous school shootings, Bernie Sanders' movement for radical reform, and the initial work of Black Lives Matter, *Avatar*, and *The Hunger Games* series seemed to blossom as a counter to right-wing paranoia and authoritarianism.

8.3 Rage and Violence in Left- and Right-wing Films, 2012–2018

These film projects, all finished or begun during Obama's first term, were soon challenged by another group of movies, however, that took their energy and orientation more directly from connective media on the Internet. In retrospect, this turning point occurred for left-wing films with the release and phenomenal success of *Django Unchained* in 2012, even though *Django* mixed a generally left-wing message with right-wing vigilantism. The transition and reorientation were slower on the right, as the techniques of animation gradually erased contextual ties to social–historical reality and muted empathic response in several animated superhero films, while also pumping up the level of personal outrage. What eventually replaced the concern with civic justice and vigilantism that centered the *Dark Knight* series were emotions and themes that had originally emerged in more mainstream, center-left films. With the release of *Deadpool* in 2016, a superhero film that embraced the ethos and behavior of authoritarian social dominators, that transition was virtually complete. I will begin with a discussion of *Django*, then look at the rise of animated violence in several mainstream superhero flics before turning to *Deadpool*. As will be clear, the popularity of rage and violence in both of these left- and right-wing films may be partly understood as the result of changes in the role of connective media in American culture and society.

Although Biskind's discussion of filmic polarization covers a lot of ground, his analysis cannot account for the surprising popularity of Tarantino's *Django Unchained* in 2012. To be fair, the director's usual combination of satire and cartoonish violence has often pulled filmgoers in contradictory ways, but *Django's* addition of Black vengeance to the mix spins the film out of Biskind's usual categories. The complicated plot involves Django, a former slave in the antebellum South, who sets out with a White partner to rescue his wife from a cruel owner on the "Candyland" plantation in Tennessee. After the plantation owner discovers the intentions of the pair and his partner is killed, Django goes on a rampage, killing a guard, the owner's lawyer, and several gunhands. Following plot complications designed to provoke even more vengeance, the "unchained" protagonist kills several more bad White guys connected to slavery, including the agents of an Australian mining company. In a fiery climax, Django sets dynamite around the Candyland mansion, kills the rest of the old folks at home, releases two house slaves, and blows up the place, as he and his wife look on from a distance.

Despite its occasional patina of historical accuracy, *Django* combined the righteous revenge at the end of *Avatar* with the superhero acrobatics of Marvel Comics movies to create a film that likely appealed to extremist elements on both the left and the right. Some left-wing spectators probably enjoyed Django's dispatch of evil Whites upholding slavery while others on the right applauded his vigilante justice; both rewarded Tarantino with the highest grossing film of his career. Even stranger, the Academy also gave him an Oscar for screenwriting this bloody mess of a plot in 2013.

Superhero violence is also in your face throughout *Deadpool* (2016), a part of Marvel Comics' *X-Men* series. Begun in 2000 to challenge DC's comic book and film supremacy—all Batman films, for example, paid DC Comics for story rights—the series hit its stride with *X-Men: First Class* (2011). The film features the X-Men team helping one of their own, Erik/Magneto, to avenge the killing of his mother during the Holocaust in Nazi-occupied Poland. The generally liberal-minded X-Men superheroes take a turn to the right to support Tony Stark's capitalist self-interest in *First Class*, however, when he refuses to turn Iron Man over to the government. Stark argues that he invented Iron Man and should be allowed to control it. In a later film in the series, *Captain America: Civil War* (2016), the Captain proclaims his independence from UN governmental supervision, despite the collateral damage that he and the other superhero Avengers are causing. After accepting the label of "vigilante," the Captain adds, "I never really fit in anywhere, even in the army. My faith's in… individuals" (quoted in Biskind 2018: 142). So much for expecting team cooperation and international justice from a formerly center-left series!

Deadpool (2016) completed Marvel's transition from the mainstream to the dark side by endorsing much of the rage and violence of authoritarianism on the right. Although, says Biskind, "it took Batman a long time to shed his code [of restraint] and embrace the beast within, [Deadpool] does not bother to burnish his bad behavior with liberal pieties like the superheroes in so many Marvel shows" (Biskind 2018: 163). Tough-guy Deadpool is leading a normal life of stealing from pizza delivery boys and falling for sexy women when he learns he has cancer and decides to fix it by getting an injection that will transform him into a superhero. After the shot, his body has bulked up but his face is blotchy and deformed.

In a plot twist that nods to Marvel Comics' primary audience of nervous teenage boys, Deadpool knows he cannot face his girlfriend (Vanessa) with a complexion that looks like a terrible case of acne (and right before the prom!), so he avoids her, puts on a superhero mask, and goes after the villainous doctor (aka Ajax) who transformed him into a monster.

Deadpool soon powers past the superslave side of the serum and much of the film features his vengeful bludgeoning of Ajax, especially after the villain kidnaps Vanessa. As in *Coxinga*, the Japanese puppet play that celebrated superhuman militarism, *Deadpool*'s animation cancels the painful empathic effects of a real-life slug fest, reducing the superhero-Ajax contest to puppet-like choreography. Deadpool kills Ajax in the end, telling one of his superhero sidekicks that "if wearing superhero tights means sparing psychopaths, then maybe I wasn't meant to wear 'em" (quoted in Biskind 2018: 164). When Deadpool removes his mask, Vanessa, however, doesn't mind his complexion. "Wow! It's a face I'd be happy to sit on," she purrs. Biskind remarks that Deadpool may be "the first alt-right superhero" (163), to which I'd add that the entire film may be understood as a recruitment tool aimed at male teens for alt-right extremism. No pain and plenty of exciting action to gain!

The rage and revenge churning *Django Unchained* and *Deadpool* are surprisingly alike, despite each film bona fides in left- and/or right-wing orientations. Anchoring both is a super-charged emotional reasoning that ignores broader problems of social justice to focus narrowly on a quest for individual revenge. Django does not justify his desire for vengeance against slave owners on the basis of historical abolitionist morality or even on a more immediate concern for the welfare of all the slaves on the Candyland plantation. He's simply enraged by the threat of rape against his wife and this rage fuels and implicitly justifies the bloodbath of the plot. Similarly in *Deadpool*, the stakes and the anger are entirely personal. Freed from the norms of UN supervision and the expectation of superhero teamwork, Deadpool is also "unchained" and can pursue his vendetta against Ajax, which results in a locker-room battle to see who gets the girl. Both heroes are also psychologically damaged. Django admits that his own years in slavery drive his obsession for revenge and the creators of Deadpool made him strangely self-conscious about his complexion and social role-playing, identity problems that mark him mentally as a teenager. Finally, a Black-and-White morality drives each plot. This is literally true in *Django*, of course, where the racial divide, following the death of Django's White partner, structures the good-guy/bad-guy divide and the killings that follow from it. Also, unlike the *Dark Knight* series or *Avatar*, moral clarity arrives early in these films. This is too bad for the mining agents from Australia, where the institution of slavery had been banned throughout the British Empire during the purported historical time of *Django*. Those White guys apparently deserved to die anyway, just for getting themselves mixed up in US slavery. Although there are fewer bodies to bury at the end of *Deadpool*, there's lots of other collateral damage to clean up. Too bad about that too, but it seems that superheroes deserve to pursue their own libertarian ends, regardless of the messy consequences.

To understand what's going on and why in *Django Unchained* and *Deadpool* I turn to *The Coddling of the American Mind*, by sociologists Greg Lukianoff and Jonathan Haidt. The authors attack "three Great Untruths," all facilitated, if not directly caused by the connective media of Internet culture—the untruths of "emotional reasoning, psychological fragility, and us-versus-them dichotomies" (Lukianoff and Haidt 2018: 4). The authors state that they first noted these changes in 2013 on college campuses, when many students began to demand restrictions on public debate on the basis of these falsehoods. The students claimed that some ideas adversely affected their feelings, leaving them traumatized and disempowered. As a result, they tried to ban speakers that made them feel this way—speakers they believed to be simply evil people. In response, note Haidt and Lukianoff, many administrations acceded to student wishes and narrowed the open expression of ideas on their campuses. From the authors' perspective, schools and universities should be challenging students with new ideas, even if they are mildly disturbing. They point out that ideas alone are not traumatizing and that students and their enablers need a more rigorous understanding of what it means to be truly disempowered. Finally, Haidt and Lukianoff note that the world is a complex place; Black-and-White morality seldom produces adequate ethical assessments, which must often bow to shades of gray. Consequently, the culture of coddling should end because it has compromised academic freedom, shrunken the arena for public debate, and perversely limited the ability of students to take charge of their own lives.

It is not difficult to link these challenges and compromises to the social pressures exerted by connective media culture. As noted in the last chapter, social psychologists have tied an increase in Internet use to higher rates of depression and anxiety, poor cognitive functioning, and a general decline in well-being among young adults. *The Coddling of the American Mind* also confirms these findings. Facebook, Snapchat, Twitter, YouTube, and other connective media platforms certainly encourage immediate, emotion-charged responses, rather than the nuanced dissection of complex arguments and positions. In the comfortable bubble of on-line "friends," it is easy to divide the world between insiders and outsiders, to engage in "virtue signaling" to affirm group loyalty, perceive "microaggressions" against allies, and "call-out" the sins of enemies. Lukianoff and Haidt analyze each of these on-line moves and their typical "cancel culture" results within the context of their three untruths. They cite the early work of Trent Eady, a Canadian queer activist, who wrote about his escape from this mindset in a 2014 essay entitled, "'Everything Is Problematic': My Journey Into the Centre of a Dark Political World and How I Escaped." According to *Coddling*, Eady identified four characteristics of this culture, "dogmatism, groupthink, a crusader mentality, and anti-intellectualism," all of which add up to the authors' three untruths. More broadly, say Haidt and Lukianoff, the groupthink of us-vs-them morality amounts to a repudiation of "common-humanity identity politics" (as affirmed by all universal religions) for "common-enemy identity politics." When joined with crusader emotions and hypersensitivity to microaggressions, this approach to social relations can produce both a loss of self-confidence and a rise in aggressive public shaming. Indeed, it is not a stretch to find both of these outcomes on display in *Django Unchained* and *Deadpool*, with crusader shaming transformed into punishing vigilante justice.

It is important to note that these Internet effects, while significant, did not transform most Hollywood films. Marvel and DC continued to produce traditional superhero pictures, including *Wonder Woman* (2017) and *Black Panther* (2018), both generally center-left and family-friendly. *Django's* category-busting success probably helped to propel the creation of more films with heroic Black protagonists, but these ranged from *Twelve Years a Slave* (2013), to *Selma* (2014), and to the magnificent *Moonlight* (2016), all of which, in surprisingly different ways, defied *Coddling's* three untruths. Although the mindset of Internet connective culture did not take over mainstream Hollywood filmmaking, it did scramble the structure and effects of dysphoric films on the left and right wings—before 2012 the pole stars of political Hollywood. With *Django* and *Deadpool*, the clear distinctions separating *The Dark Knight* trilogy from *The Hunger Games* series dissolved and converged into a single nexus that emphasized super-charged rage, common-enemy identity politics, and grievances that motivated personal revenge. These social problems would drive several more movies and a significant series for the next five years, right into the middle of Trump's presidency.

One of the most dysphoric was a dystopian horror fantasy called *Get Out* (2017), another film like *Django* about White people behaving badly who deserve to be killed by a heroic Black man. A group of elite White folks entrap young Black people and then implant the brains of White grandmothers and grandfathers into their bodies so that the old folks can turn these Black bodies into their own avatars and thereby extend their lives of White entitlement for another fifty years or so. Whereas Cameron imagined his avatars as an image of cross-racial possibility, the writers of *Get Out* reimposed the color line through a literal Cartesian mind–body separation in order to segregate the races and signal the destruction of Black culture by predatory Whiteness. The film focuses on a smart Black guy who initially falls for a White girl, figures out the switcheroo that the girl's family plans as his fate, correctly identifies it as a new form of slavery, kills the family (including his former girlfriend), and, with the help of an African American computer geek, turns the rest of the racist suburban community over to the cops. As in *Django*, common-enemy identity politics structures the film's morality and righteous crusader revenge powers the resolution of the plot. The film has the feel of left-wing radicalism, but its politics are closer to Internet groupthink than cooperation, egalitarianism, and xenophilia.

One of the most confused—and therefore, revealing—film and television series to result from the emotional reasoning and virtue signaling of connective media effects has been *The Purge* (2013–2018), which began as a horror film, quickly expanded into a TV series, and switched politics to generate two more films. The initial 2013 film probably found more fans on the right than the left. Through a voiceover at the start of the film, spectators are told that all crime in the dystopian America of 2022 is now legal once a year, when citizens are allowed to purge their emotions through violence during a 12-hour period. The audience also learns that these acts of murderous catharsis have become an accepted method of population control, with the result that crime and unemployment are now down and the economy is booming after the elimination of so many poor people of color. The social-historical assumptions behind this given set of circumstances already mark the film as an authoritarian fantasy.

Like many horror films, most of *The Purge* occurs in an enclosed space, the home of the Sandin's, which the right-wing patriarch of this nuclear family has barricaded with armor—he's in the business of selling home protection equipment—in preparation for Purge Night. Before lockdown inside the house, however, the pre-teen Sandin son sees a poorly dressed Black man looking for a place to hide in their wealthy neighborhood and invites him in; the stranger rolls under the armored door just before its final descent. Meanwhile, teen daughter Sandin is hiding her boyfriend in her room and he tells her that he needs to talk with her father about their relationship. But instead of a conversation, the boyfriend pulls a gun and Dad must kill him to avoid his own death. Right after that, Dad discovers the strange Black man in the house, but the Stranger escapes and successfully hides in a back compartment of the son's upstairs closet. All of this occurs in the first fifteen minutes of the film and it's already apparent that patriarch Dad will have his hands full on Purge Night.

Next, a masked gang of young criminals and neighbors begins destroying the barricades around the Sandin home, vowing to avenge the price-gouging patriarch for overcharging them for their home protection systems and also looking to kill the Black Stranger that one of them saw entering the Sandin home. The horror starts soon after, as a Halloween-masked gang, armed with guns, knives, axes, and implements of torture, tries to kill the family, which now cannot escape from their mostly sealed home. Dad dispatches many of them, but the leader of the gang kills him as more neighbors, which include some multicultural pairings, close in on the remaining Sandin family members at the dining room table. They are about to kill Mom, Daughter, and the young Son when the Black Stranger appears behind the leader of the gang and threatens to kill him if anyone touches the Sandins. Mom, who has gained courage throughout the night, tells everyone to remain where they are until Purge Night is over, which she believes will end the crisis. They do, and the camera pulls back to dwell on this odd tableau: Black Stranger standing behind the White Enemy of the Family, knife at his throat, Daughter and Son seated on either side of Mom, in the center, and multicultural couples seated and standing around them. Only Dad, now dead in a corner, is missing—and perhaps a plump Thanksgiving turkey—to complete this Norman Rockwell-like family portrait.

What to make of *The Purge*? Although grotesque satire was probably intended to crown its ending, most of the film works as a stock horror movie, with all of the chills and thrills of surprise violence that the genre was built to deliver. The Black Stranger primarily functions as an element of danger during the movie and Dad, despite his miscalculations and patriarchal foolishness, does save his family from the murderous violence of the neighbors to become something of an altruistic hero. In terms of the film's orientation toward human nature, there is no doubt that purge life is closer to Hobbes than Rousseau—mostly "nasty, brutish, and short" and requiring right-wing authority and masculine strength for survival. In this regard, much of the film could easily function as an advertisement for the NRA; yes, guns can kill, but you've got to protect yourself on Purge Night. Given Tuschman's understanding of the left's take on hierarchy and tribalism, however, it is also evident that poor, non-White strangers may be necessary to save rich White families, especially when the envious try to kill them. Violent rich people, in fact, may be more of a problem than racial and ethnic otherness, which may explain why the well-dressed Hispanic, Asian, and ethnically mixed couples at the end continue to threaten murder against the remains of the Sandin family. But whether most spectators understood this as the likely meaning of the final tableau or took away another message entirely cannot be known. Few horror films attempt satire at their end.

The politics of *The Purge* sequels, however, are much less ambivalent. *The Purge: Anarchy* (2014) and *The Purge: Election Year* (2016) feature-poor Black and Hispanic characters who band together in an anti-Purge resistance group to save a working-class mother and daughter (in *Anarchy*) and help to protect a White female candidate for President during Purge Night (in *Election Year*). In that sequel, the mostly male, rich White oligarchs who now run the US government kidnap the Senator running for President—a sexier, younger version of Hillary Clinton in 2016—and attempt to wipe out the anti-Purge resistance groups. *Election Year* climaxes in a *Godfather*-like shootout, Tommy Guns Blazing, that kills more evil White men than *Django Unchained* and saves the Senator for her (presumed) election to the presidency. By 2016, *The Purge* franchise, TV series included, had embraced the patina of left-wing politics, but remained centered on conspiracy theories, common-enemy morality, and tribalistic vengeance.

Why encourage extremists to hope for a Purge of right-wing capitalists and autocrats before the next election? This emotional reasoning may be a pleasant daydream for some connective media wannabe lefties, but it places Black and Brown bodies on the front lines of opposition in a potentially bloody conflict with the US troops and police that would certainly be marshalled to defend the status quo. This vision of radical rebellion ignores the reality of White supremacy embedded in American history since 1619. But this Internet-fed Purge fantasy avoids such hard historical lessons to jump to shoot-'em-up conspiratorial solutions. The depressing fact of the entire Purge series is not that Hollywood would try to cram so much nonsensical horror into its dysphoric entertainment, but that so many Americans apparently found the nihilistic and contradictory series believable for five years.

Universal Pictures began *The Purge* series with star Ethan Hawke and high production values, but its TV part of the series has now been cancelled and its last film, set to be released in 2021, remains TBA. No such diminished popularity greeted the release of *Joker* in 2019, however—a film that shares much of the ambivalence of *The Purge*. Billed as a psychological thriller, *Joker* is set in the early 1980s in the realistic urban decay of Gotham, which encourages the audience to understand Joaquin Phoenix's character, Arthur Fleck, as a kind of prequel to explain the origin of the Joker character in the *Dark Knight* Batman series. Arthur suffers from a disorder that causes him to laugh at inappropriate times, but he is struggling to overcome this disability at his job as a clown amusing kids in a hospital, patiently dating a single woman his own age, and attempting to start a career as a stand-up comic. His life starts to fall apart, however, when he is unfairly fired from his job and beaten up by three rich businessmen on a subway. Enraged, Arthur, still in his clown costume from work, shoots two of them in self-defense on the train and chases down the third, whom he executes. His vigilante killing inspires admiration as well as a police investigation and soon street demonstrators are appearing dressed in clown costumes to protest their treatment by the rich. Funding cuts in Gotham stop Arthur's meds and he spirals downward, shooting others to cover up the initial killings. Intrigued by clips of Arthur's stand-up routine and the notoriety of the Joker-clown image, TV personality Murray Franklin, played by Robert DeNiro, invites him on his comedy show and Arthur appears in his costume to audience applause. Soon, though, after telling morbid jokes and confessing to the subway killings, Arthur shoots Murray on-air for criticizing him and is quickly arrested. But rioters dressed as jokers crash into the police car driving to the station, freeing Arthur, who stands on the car, dances as the Joker to the music of their rioting, and smears his lips with blood to create his trademark grotesque smile. The last scene shows Arthur in a mental hospital running away from orderlies and leaving a trail of blood behind him.

Conceived and written by director/co-writer Todd Phillips in 2016 and 2017, *Joker* was an enormous US and international hit. After its release in August, the film grossed over a billion dollars by the end of 2019, the only R-rated film ever to reach that total in under a year. *Joker* earned eleven Oscar nominations and won two of them, for Original Score and Best Actor, making Phoenix the second actor after Heath Ledger to win an Oscar for playing the Joker. The Golden Globe and BAFTA ceremonies also awarded Phoenix and the composer with top honors. Todd Phillips drew on a 1988 graphic novel, *Batman: The Killing Joke*, but admitted that he was primarily inspired by two films directed by Martin Scorsese, *Taxi Driver* (1976) and *The King of Comedy* (1982). DeNiro starred in both—as Travis Bickle, the alienated and eventually homicidal taxi driver in '76 and as a pathetic stand-up comic, angling for success in '82. Not surprisingly, Phillips went out of his way to hire DeNiro for Murray Franklin, a kind of reprise of his earlier role in *King of Comedy*.

Regarding euphoric and dysphoric elements in the film, *Joker* is nearly devoid of euphoric incidents and emotions; in fact, the film is likely the most dysphoric drama I have analyzed for this book. Despite its urban terrors, the *Dark Knight* series held out the hope that justice, whether through police reform or Batman's vigilante heroics, might win out in the end. *Django* and *Get Out* embraced vigilantism, but at least acknowledged the racism and oppression of American history. Even the first *Purge* kept Mom and the kids alive at the end in the expectation of returning to some kind of normal after Purge Night. In *Joker*, however, scenes centered on deprivation, fear, humiliation, disgust, pain, and murder— all dysphoric elements—are unrelenting and, finally, overwhelming. Even worse, attempts to lighten the mood through an occasional injection of euphoria, such as stand-up jokes from Arthur, typically turn grotesque. His misplaced laughter echoes throughout the film and the eventual smile painted on his lips drips with blood.

Released to a polarized audience in 2019, *Joker* also polarized the critics. Although one reviewer, Chauncey K. Robinson, noted that "the film walks a fine line between [the] exploration and [the] validation" of Joker's character, most critics and some film directors quickly found themselves on one side or the other of that line (*Joker* [2019 film] 2020: *Wikipedia* 12–13). A British neurocriminologist, for example, was impressed by the film's accurate depiction of the psyche of a murderer and an American psychiatrist noted that poor people like Arthur with mental illness have no chance of getting help in the morass of the American health care system. Their sympathetic concern for the Arthurs/Jokers of the urban scene was shared by journalist-reviewers for *The Guardian* and *The Washington Examiner*. Director Michael Moore went so far as to call *Joker* a "cinematic masterpiece" and said it would be a "danger to society" if people did not see it (2020: 12).

On the other side of the line, several writers expressed the belief that *Joker* might inspire actual violence. The reviewer for *Vanity Fair* found the film too sympathetic toward "White men who commit heinous crimes" and descried the "woebegone mythos" placed on those who "shoot up schools, and concerts, and churches" (2020: 13). Similarly, the *National Review's* critic worried "that a certain segment of America's angry, paranoid, emotionally unstable young men will watch Joaquin Phoenix descending into madness and a desire to get back at society by hurting as many people as possible and exclaim, 'finally, somebody understands me!'" (13). Evidently fearing such a response, the owner of a movie theater in Aurora, Colorado, where a mass shooting occurred during the viewing of *The Dark Knight Rises* in 2012, refused to screen *Joker* in 2019.

In terms of social cohesion, however, these two critical camps, despite their opposition, arrive at the same conclusion. As noted, Whitehouse is concerned with actions and situations that bind groups of people together, such as honor codes, oath-taking, family alliances, and common-enemy affiliations. Arthur/Joker has strong ties to no social groups and becomes even more alienated and isolated as the film proceeds. He turns away from his sometime girlfriend and mother and eventually kills a couple of his co-workers from the hospital because they know where he got the gun he used to shoot the three businessmen. As a "domestic terrorist," Arthur fits the profile of the "lone gunman" who attacks others because his grievances overwhelm him and he is desperate to draw attention to his own needs. But social cohesion in *Joker* has to do, as well, with the potential ties that might bind the city of Gotham together. The underlying question here is whether Gotham—aka, New York, LA, Philadelphia, Chicago, etc.,—is worth saving. To that important question, both groups of critics implicitly answer "no." Gotham is a terrible place for the mentally ill, says one group, and the other group adds that films like *Joker* will only empower more psychopaths and make Gotham worse. Of course some critics would also suggest possibilities for urban reform or even radical Reconstruction, but the film itself has no heroic social workers or even billionaire businessmen offering to build trust and promote cooperation across class and tribal lines. By the end of the film, Gotham is filling up with Jokers and there is no one around to drain the cinematic swamp.

Significantly, the ambivalent critical response to *Joker* contrasts sharply with the praise generated by *Taxi Driver* in 1976, the film that it most resembles. Most critics celebrated director Scorsese and DeNiro's success in depicting a troubled young man driven toward extreme actions by the circumstances of his life. The clear markers of increasing psychological distress provided by Bickle's relationship with two women marked his descent into murderous violence. In brief, the screenplay, together with some excellent acting and directing, encouraged the audience to adopt the point of view of the women who were trying to figure out what DeNiro's Bickle was up to. By sympathizing with two other characters who empathize with Bickle, we see DeNiro's character through the eyes of Betsy (Cybill Shepard) and Iris (Jody Foster). Early in the film, Betsy is initially attracted to Bickle, then repulsed when he takes her to see a pornographic movie. Bickle's failure with Betsy leads him to plan to kill the politician on whose campaign she is working, but he backs off when a secret service officer approaches him. The next woman is 14-year old Iris (Jody Foster), a run-away teen trapped in an abusive sexual relationship with a Black pimp. When even Iris begins to reject Travis after he offers to save her, we understand that nothing good can come of Bickle's alienation. He kills her pimp and his compatriots, saving Iris at the climax of the film. But Scorsese treats the rescue scene and Bickle's celebration as a hero by the police and the public ironically, underlining the craziness of his attempt at vigilante justice. No doubt part of the response in 2020 to Bickle's murderous heroics at the end would include a critique of his White (vigilante) privilege, but Scorsese's use of secondary empathy and irony also open up the need to see the problems of Gotham through fresh eyes.

In contrast, Todd Phillips, both in his script and in his directing, avoids such secondary empathy and the irony that can complement it. The audience soon learns that they can trust no other character's reactions to Arthur; in fact, his psychological fragility often keeps him so isolated that it is difficult for spectators to fully comprehend what he's doing and why. Further, Arthur's problems and emotional obsessions soon crowd out other points of view, creating a claustrophobia that makes irony impossible. Instead of secondary empathy and irony, the Arthur-vs-the-world organization of the story pummels the audience with fear, shock, and pity. Lukianoff and Haidt's "three untruths" not only muddy Arthur's psychology but are translated into plot episodes that drive the film to multiple murders and the complete breakdown of public order in Gotham. Spectators are left with no hope for alienated and disturbed urbanites like Arthur Fleck.

In this regard, the connective media of the Internet have apparently impeded the communicative operations of what Whitehouse terms orthodoxy and orthopraxy for *Joker*. If, following Whitehouse, we understand the primary function of media critics as delivering orthodox interpretations of the meanings and relevance of *Joker*, it is apparent that the film sent different groups of Hollywood high priests in two opposite directions. While the rival interpretations may lead to few real-world differences for the social cohesion of actual Gothams, the fact that the critics were so divided suggests that the polarization tearing apart the polity is also dividing popular critical discourse. Regarding orthopraxy – social checks to make sure that rituals are performed in a standardized way—it is clear that although Todd Phillips may admire Scorsese's films, he has very different ideas about the ingredients of a good movie. Scorsese developed his directorial artistry years before Internet communication challenged several of the givens of filmic communication. To judge from the accolades won by *Joker*, Phillips is responding to new sensibilities that privilege psychological fragility, emotional reasoning, virtue signaling, and common-enemy identity politics. Needless to say, this is not an orthopraxy designed to increase the social cohesion of the nation.

Pressed by the spreading pandemic, Hollywood stopped making and releasing all but the most mainstream, potentially profitable films during 2020. At this writing, the industry's response to extremist, dysphoric pressures from the polarities of the right and left is unclear.

8.4 Trumpian Carnage in the Time of COVID-19

During the first three years of the Trump administration, critics typically breathed a sigh of relief that the nation and the President had not yet faced a serious crisis. We can get through four years of Trump, many believed, and still return the country to something like normal after the Democrats defeat him in 2020, if nothing terrible happens. As I write this final section of Chapter 7 in January of 2021, it is evident that the effects of the pandemic and the Trump administration's response to it on social cohesion in the US have demolished that modest expectation. At the same time, however, it may be that the possibility for the beginnings of a reversal of present social and political trends has arrived at a turning point that could provide the basis for rebuilding US social cohesion on the basis of some guarded coevolutionary optimism.

Many have noted that the pandemic has torn away the now shredded band-aids covering numerous social and economic wounds suffered by the US body politic over the last fifty years. Indeed, it is difficult to imagine a more revealing blow to the assumptions of neoliberalism than a pandemic—especially the belief that government should keep its hands off the marketplace and support more wealth and power for the rich. Although many citizens expect their local, state, and federal leaders to provide truthful information, to pass necessary laws for the common good, and to provide emergency health care, assistance, and funding to help as many as possible survive the ravages of the disease and its numerous social ramifications, neoliberalism's cupboard of tricks to restore trust in government is bare. Accordingly, this section will examine how Trump and his followers responded to the multiple crises of the pandemic, the ethical and political problems this created for the President, the damage to the social cohesion of the nation, and some of the work that remains to be done to reverse the present trends toward national disintegration.

First, it is evident that Trump understood the pandemic primarily as a political challenge to his reelection and not as a potential medical and humanitarian catastrophe. Although the President was told in January of 2020 that COVID-19 could quickly mushroom out of control, kill thousands of citizens, and wreak havoc on businesses and communities, he chose to downplay the threat to the public so that the American economy—which he viewed as his best ticket to reelection—could continue to generate jobs and profits. Even though that strategy failed in April during the first wave of the pandemic, when unemployment rose to around fifteen percent, Trump never abandoned it. Following his usual authoritarian playbook, he simply looked for scapegoats and found them in China, the World Health Organization, "deep state" conspiracies, and Democratic state governors. Pretending that COVID-19 was not very contagious or deadly allowed him to parade as the protector of jobs and the economy. The "Liberate Michigan" and "Liberate Virginia" slogans that came out of his tweets and rallies became wedge issues, separating his MAGA supporters from the Dems and promoting common-enemy identity politics and vigilante justice. Likewise, Trump's refusal to wear a mask rallied macho social dominators as well as conforming authoritarian followers to his cause. Never mind that all of these campaign strategies also spread the virus and no doubt contributed to the deaths of thousands.

Nor did it matter to Trump and his followers that his campaign of deliberate confusion put the administration at cross purposes with scientific evidence and good medical practice. As he had done at the Environmental Protection Agency and elsewhere, Trump appointed political loyalists to run or oversee the Food and Drug Administration and the Centers for Disease Control. Although the CDC stumbled badly in its initial attempt to put together a reliable test for Covid-19, US public health agencies began 2020 with some of the best medical expertise in the world. But Trump strove to control their messaging about testing, mask-wearing, school closings, and state and local regulations, modifying and muddying crucial guidelines that could have saved many lives. Trump loyalists also prevented the Occupational Safety and Health Administration from intervening in situations that could have prevented thousands of COVID-19 deaths at meat packing factories across the nation. As investigative journalist Jane Mayer reported, one spokesperson for the legal rights of the many Black and Hispanic laborers in the meat and poultry processing plants concluded that "the industry is getting away with murdering people" (Mayer 2020: 32).

Of course putting his political priorities above the rights and lives of minority workers would not have been a problem for his authoritarian followers, if they ever bothered to find out about it. But there is reason to believe that many citizens were troubled by Trump's indifference to the human suffering caused by the pandemic. Altruistic acts and gratitude toward those who perform them are frequent responses to calamities and emergencies. Typically, they lead many to express heightened feelings of humanitarianism and solidarity across all afflicted groups that may otherwise be indifferent or even opposed to each other. A poll of 8,000 Americans conducted in March of 2020 by the Centre for Experimental Research on Fairness, Inequality, and Rationality (FAIR) found that the pandemic "has increased American's feelings of solidarity with others," which may, "over time, affect public opinion on lessoning the social and economic impact of the virus" (Cappelen et al. 2020: April

19). They added that "these changes may have important implications for politics. The study shows that respondents who prioritized America's problems over their own were more likely to favor economic redistribution and universal health care" (2020: April 19). But other results from FAIR's polling contradicted these conclusions. When they thought about the pandemic, some Americans concluded that contracting the virus was largely a matter of good or bad luck. From this, they concluded that because luck plays a major role in life, there is little point in working to mitigate the pandemic and its effects on racism and inequality. Taking both findings together in mid-April, the researchers stated: "For the moment, the survey suggests that the shifts are effectively cancelling out each other, leaving overall support for policies such as universal health care unchanged" (April 19).

In a piece for the Sunday *New York Times*, political philosopher Michael Sandel provided one answer to the FAIR findings. Recognizing that the pandemic has excited strong feelings of both fatalism and solidarity, he urged that citizens concerned about the future of democracy in the US should "pursue the intimations of solidarity implicit in this moment to reframe the terms of public discourse, to find our way to a morally more robust political debate than the rancorous one we have" (Sandel 2020: April 19). Indeed, this was the eventual response of the Biden campaign to Trump's combination of cynicism, lies, and indifference concerning the suffering caused by COVID-19. Although the connection between the pandemic and Biden's eventual victory has not been rigorously studied, Sandel's article suggests a likely logical link.

It would take another month of rising pandemic deaths, armed right-wing protests against the shut-downs, and heightened economic anxieties before Sandel's plea for a morally robust political debate built upon altruistic solidarity began to occur. Unfortunately, it also took the murder of George Floyd by a Minneapolis policeman and his three accomplices in blue to break the dam of unrelieved grief and pain resulting from the pandemic and focus public attention on police brutality. Floyd's killing followed soon after the deaths of two other Black victims of White violence that were also in the news. The video of the May 25 strangulation, widely broadcast for two weeks after the incident, showed a cop nonchalantly kneeling on Floyd's neck for eight minutes and forty-six seconds—an interval that would live in infamy—while Floyd could be heard saying "I can't breathe" as the other policemen held him down and kept protesting onlookers from intervening. The initial demonstrations targeting police brutality, led by Black Lives Matter (BLM) and allied progressive groups, soon forced the dismissal and arrest of the police involved. They have since been charged with murder and accomplice to murder.

But, as in Fergusson, Baltimore, and several other cities that BLM targeted in the protests of 2012–2015, looters, arsonists, radical anarchists, and White supremacists soon joined the overwhelmingly peaceful demonstrators to push their own desires and agendas. Would White citizens understand the protests against Floyd's killing as a righteous demand for justice or an unnecessary, even riotous challenge to normal law and order? For a week or so after the strangulation, as television cameras showed both looters and arsonists as well as nonviolent protesters in Minneapolis, Nashville, Washington, DC, and many other cities, it looked like that response could go either way.

There were several reasons why most White Americans tentatively sided with the peaceful protesters by mid-summer. Although the widespread outpourings of rage against Floyd's murder were initially compared to the situation in American cities in 1968 following the assassination of Martin Luther King, the crowds were different and many more city mayors and governors professed support for the protestors and worked to quell the outbreaks of pillage and arson that accompanied some of the marches. As well, BLM leaders, realizing that looters and right-wing vigilantes could hijack television and social media commentary, also worked to restrain expressions of violence. Finally, revulsion against Trump's response in DC to the protests may have tipped the balance away from tribalism and toward altruism and solidarity. A week after the killing in Minneapolis, President Trump sent marching and mounted federal troops, armed with tear gas and pepper bullets, plus a low-flying helicopter, against peaceful demonstrators in DC, near Lafayette Park and the White House. Trump's authoritarian and legally dubious actions probably alienated more Americans than they convinced and also animated many more demonstrators across the nation to join BLM in the streets for another week of protests. A poll taken soon after the incident showed that over 70% of White Americans at the time believed that racial discrimination was a big problem.

Disgust with Trump's racist actions was another reason, then, for BLM's success in harnessing the outpourings of rage into protests against police brutality. By this time, it was also clear that COVID-19 was much more dangerous to Black citizens than White ones and many leaders began calling for better statistics to track the problem. Because Black Americans suffer from higher rates of diabetes, hypertension, obesity, heart disease, and asthma—and because they are more likely to experience poor schools, bad housing, polluted air and water, and other damaging environmental factors—"predominately Black counties [have seen] three times the rate of infection, and six times the rate of Covid-related death, as white counties," according to one study in early June. (Williams and Sanchez 2020: June 5). Initially it had seemed to many that catching COVID-19 was largely a matter of chance. For many citizens watching reliable news coverage, however, the impact of the pandemic on Black deaths provided another reason for despising Trump and linking support of BLM to their personal politics. Because COVID-19 attacks the lungs, "I can't breathe" became a phrase that linked the pandemic with Trump's racism. And Trump himself was already understood by many as a leader who could express no genuine empathy or sympathy with the victims or heroes of the pandemic.

For these reasons, former Vice President Biden's presidential campaign to "restore the soul" of America gained credibility with thousands of citizens who might otherwise have stayed home or even voted for Trump. Of course there were differences in policy as well that may have prompted poor and working-class voters to opt for the Democratic candidate— health care, infrastructure, racial justice, immigration, an increase in the minimum wage, and protection from the ravages of climate change—but the Democrats may not have won the presidency on matters of policy alone. When all of the votes were tallied, Biden defeated Trump by seven million votes in the largest voter turnout in US election history. Feelings of altruistic solidarity likely rallied voters on both sides, with Trump even increasing his votes by four million above his 2016 totals. While both campaigns can take some credit for the increase, it is likely that anxiety and grief over the pandemic and the resultant altruism about standing with your in-group helped to prompt the huge turnout. In the electoral college, Biden's margin of victory was 306 to 232, but had a mere 65,000 votes shifted in key states and districts, Trump could have won the election, despite Biden's overwhelming majority of the popular vote.

The 2020 election was one of the few in recent memory when neither presidential candidate had substantial "coattails," the influence that the popularity of top-of-the-ballot candidates can exert on voting for candidates in the same party below them. Despite Biden's majority, the Democrats actually lost several contests in the House and their two Senate victories probably had more to do with the candidates they ran than with Biden's popularity in Arizona and Colorado. Most Republicans who won races in the House actually ran ahead of the President's numbers in their districts; they were more popular than Trump was. The lack of coattails for both candidates certainly substantiates the continuing reality of massive political polarization across the country. But it also suggests that moral revulsion against Trump's personality, not necessarily his policies (which were widely embraced by Republicans running for office) was a factor for some White voters. Another is that Biden's old-fashioned morality honoring empathy and decency worked for even more voters than did Trump's implicit pitch for neoliberal selfishness and national cynicism.

Whatever the reasons, Trump is leaving President Biden a nation that is more politically divided and closer to social and political disintegration than at any time since the Civil War. Given Trump's refusal to concede that he lost the election, one immediate problem for Biden and his team will be the legitimacy and competence of government itself. As Turchin remarks in *Ages of Discord*, "Probably the most direct indicator of state legitimacy in America is provided by polls asking, 'How much of the time do you trust the government in Washington?'" (Turchin 2016: 106–107). Before its swift decline after 1965, trust in the federal government (as indicted by combining the "just about always" and "most of the time" replies), peaked at 74%. By 2013, Turchin reports, that response had declined to 26% of the population. And by May of 2019, the *New York Times* reported that "just 17% of Americans trusted the federal government to do the right thing 'just about always' or 'most of the time'"(Tavernise 2020: May 24).

In the midst of the first wave of COVID-19, *Washington Post* columnist Dan Balz emphasized that the recent lack of public trust in the federal government simply reflected its increasing incompetence: "The government's halting response to the coronavirus pandemic represents the culmination of chronic structural weaknesses, years of underinvestment, and political rhetoric that has undermined the public trust – conditions compounded by President Trump's open hostility to a federal bureaucracy that has been called upon to manage the crisis" (Balz 2020:May 17). Besides the pandemic, Balz pointed out that the Government Accountability Office has warned about numerous bureaucratic logjams for years, sometimes for decades, to little avail. He quoted an official at the GAO who told him that more than half of the federal agencies it oversees are now at "High Risk" of failure, according to the GAO's criteria for accountability, which includes such factors as upgraded technology, relevant expertise, and responsiveness to the public. Among the most difficult federal bureaucracies to manage are those involving public health, which Balz described as "a labyrinth of government, private insurers, public and private hospitals,... [and] health-care workers, all involved in the delivery and billing of services" (2020: May 17). Balz concluded, "It is an open question whether the more intense focus on the federal government will result in more calls to deal with the underlying weakness or whether criticism of the administration's response – and the political divisions surrounding it – will further degrade people's trust in the institutions they have turned to in this moment" (May 17).

By the end of 2020, the progressive Democratic To Do list was a long one. At the top of the agenda had to be controlling and finally ending the pandemic, along with more immediate relief in food, rent, and income for the ten percent or more of the population that the pandemic had rendered unemployed and who were scrambling to keep their lives together. The list also included reforms aimed at beginning to dismantle the neoliberal economy and culture of the past 40 years that Republicans and many Democratic enablers had fostered and allowed to oppress most minorities, workers, and the poor: legislation to facilitate union organizing and gun control, to mitigate climate change, dismantle monopolies, improve health care, and restrict the movement of wealth to off-shore tax havens. Many Democrats also knew they must pass legislation to reign in the excesses of connective media and its corrosive effects on democratic norms and practices, but how to accomplish such aims remains an open question.

Some progressive goals, however, could only be facilitated through constitutional amendments. Although the Radical Republicans passed the Fourteenth Amendment primarily to protect the rights of newly enfranchised Black citizens after the Civil War, Supreme Court rulings perverted the "original intent" of the Amendment to expand the rights of corporations over people. This interpretation of the Fourteenth, in fact, was part of the reasoning behind the Citizens United decision that stripped Congress of much of its ability to restrict campaign contributions from corporations. Because corporate wealth has so corrupted US elections, reversing the Citizens United decision may be the single most important change that Biden's administration should push for. A close second and third would be statehood for Puerto Rico and DC and the abolition of the Electoral College to facilitate the popular election of the President. And finally, some progressives talked about reconfiguring state representation in the US Senate on the basis of population.

8.5 Insurrection and Beyond

By 2021 it was evident to many that Trump would leave office and Biden would become President on January 20. The President, however—despite losses in over sixty court cases challenging the election results—had refused to concede, continued to claim that the election was rigged, and was demanding changes in voting totals from some state officials and leading his authoritarian followers in boisterous "stop-the-steal" rallies. On January 6, President Trump incited a mob to mount an insurrection against the US government. Evidently hoping to create circumstances that would allow Trump to stay in office, the rioters broke into the Capitol and stopped House and Senate members from validating the electoral votes that confirmed Biden's election. At this writing, the evidence suggests that the Capitol Police and their local reinforcements had plenty of warning that violence might occur on the 6th of January and many questions remain about why they were so ill-prepared for the roughly 2,000 rioters that overwhelmed them.

Since the mob violence, the FBI and other agencies have been examining videotapes, conducting interviews, and gathering other evidence in preparation for legal action against the instigators and insurrectionists. The publicly available evidence, much of it shot by the rioters and posted on connective media as well as by network news cameramen, makes it clear that Trump ginned up the crowd of mostly White males with the lie about his stolen election at a rally in a park near the White House. Those who followed Trump's direction to march to the Capitol included armed White supremacists and anti-government militia members, active and retired police and military, QAnon believers, and other as yet unidentified groups of rioters. Many were costumed in hypermasculine attire and uttered slogans that justified their quest for vigilante justice with authoritarian action; their emotions reminded me of the mix of joy and rage on the faces of the street rioters at the end of *Joker*. Moreover, some rioters came prepared to trash the place and take prisoners, carrying crowbars, pipe bombs, knives, plastic zip ties (to be used as handcuffs), and even revolvers, despite a ban on such weapons on the Capitol grounds. In addition to pro-Trump flags and banners, several held signs demanding "military tribunals," "arrest Congress," and "Hang Mike Pence," who had told Trump the previous day that he had no constitutional authority to stop the counting of the electoral votes. The FBI noted that dozens of the over 300 rioters they were intending to prosecute were on the FBI's Terrorist Watch List. News reports suggested that many of these groups had likely conspired together to organize the insurrection. It was also evident that many of the insurrectionists knew where to go in the maze of the Capitol after forcing their way in, raising the question that they may have had help before and during the riot from some police, congressional staffers, and perhaps even members of Congress.

After watching the insurrection unfold on television for nearly four hours and doing nothing to call in reinforcements for the embattled police, Trump delivered a brief TV address when it was clear his coup would not succeed and told his supporters to go home. The next day Twitter and Facebook shut down his accounts for provoking violence, followed the next week by the rest of the major connective media companies. Two cabinet secretaries, several members of Tump's White House staff, and other officials in top administrative posts resigned over the next few days. The PGA association cancelled plans to hold a golf tournament at one of Trump's properties and Deutsche Bank announced it would no longer finance his businesses.

A week after Trump's attempted coup d'état, which resulted in five deaths and numerous injuries to the police, Democratic House members, along with ten Republicans, voted unanimously to impeach Trump, making him the first President in history to be impeached twice. The single article of impeachment detailed the President's "High Crimes and Misdemeanors." (See Resolution. 2021: *The Washington Post*, January 14, for the passages below.) Trump had violated his oath of office "to preserve, provide, protect and defend the Constitution" and, more specifically, must be removed because of Section 3 of the Fourteenth Amendment, which prohibits any person from holding office who has "engaged in insurrection or rebellion against the United States." In support of these charges, the Resolution noted that "Trump had repeatedly issued false statements asserting that the Presidential election results were the product of widespread fraud and should not be accepted by the American people." On the day of the insurrection, the Resolution quoted Trump urging on his followers with statements "such as: 'if you don't fight like hell you're not going to have a country anymore.'" Also included in the Resolution was a reference to a phone call that Trump made to the secretary of state of Georgia demanding that he "find" enough votes "to overturn the Georgia Presidential election" and threatening him "if he failed to do so." "In all of this," the Resolution's charges concluded, "President Trump gravely endangered the security of the United States and its institutions of Government. He threatened the integrity of the democratic system, interfered with the peaceful transition of power, and imperiled a coequal branch of government. He thereby betrayed his trust as President, to the manifest injury of the people of the United States."

One of the initial reasons for promptly removing Trump from the presidency right after the insurrection was the fear that he would cause massive damage to the nation and perhaps the world—he still controlled the military codes to set off a nuclear catastrophe—during his remaining days in office. With only a week left as President, however, the isolated commander in chief began to plan how to defend himself before the Senate with a small and shrinking staff and with the knowledge that, if convicted, he would be forever barred from running again for office. For the moment, impeachment by the House and action by the major Internet companies had contained most of the damage that the aggrieved President might cause. With ever more rats fleeing the sinking ship of Trump's brand, the isolated President was reportedly feeling desperate, angry, and vengeful.

Soon after the impeachment vote in the House, an increasing number of major corporate funders announced they would no longer be giving to the Republican Party. In response, Leader Mitch McConnell consulted with Republican funders and said he would keep an "open mind" about whether or not to vote to convict Trump at the end of the Senate trial. Other Republican Senators weighed their options as the GOP began to split into factions. Republicans know that Fox New and most of the other organs of Trumpist propaganda will likely continue to follow the party line about the election and Trump's lies. On the other side of the equation, they also realize that contributions to their political action committees are likely to shrink, at least in the short term, as many formerly reliable funders continue to turn up their noses at Trump's stench. Their main political problem is the future of the Republican voter. At the moment, in the middle of January, recent polling suggests that perhaps seventy percent of the Republicans who voted for Trump continue to believe, against all of the reasonable evidence, that the

Democrats, with the aid of Black voters in major cities, stole the election from Trump. What the Senators cannot fathom with any certainty, however, is how many voters in their state will continue to proclaim their loyalty to Trumpism by affirming their belief in his Big Lie and vote accordingly in the next primary election and beyond.

But there is also an inescapable ethical dimension to Republican calculation and this reckoning makes the coming vote to acquit or condemn Trump in the Senate trial both difficult for many Republicans and necessary for the future of democracy in the US. When Republican Representative Liz Cheney of Wyoming voted with the Democrats to impeach Trump in the House, she implicitly acknowledged the importance of joining ethical responsibility to political calculation. "The President of the United States summoned this mob, assembled the mob, and lit the flame of this attack. Everything that followed was his doing.... There has never been a greater betrayal by a President of the United States of his office and his oath to the Constitution," she said (quoted by E. J. Dionne 2021: January 14). Cheney is the number three Republican in the House and her colleagues on both sides of the aisle know that her ethical stand involves political risk. The daughter of the former vice president is already facing a primary challenge in 2022 and, even if reelected by Wyoming voters, could lose her seniority among minority Republicans in the House in a Trumpist sweep. Cheney probably reasoned that elevating her loyalty to democracy above her loyalty to Trump's Big Lie, although politically risky, was both responsible and politically feasible. Will small "d" democrats in both parties be able to frame the Senate vote as a necessary reckoning for the future of democracy, so that more Republican Senators might consider joining the Democrats in convicting Trumpism along with Trump in the impeachment trial? The situation is reminiscent of the problem for Northern Democrats in Spielberg's *Lincoln*, when the President needed a few Democratic votes to pass the Thirteenth Amendment.

If such a reckoning is avoided and Trump gets off (again), his Big Lie about the election will likely grow into another Lost Cause myth. The Confederate myth of the Lost Cause emerged after the Civil War primarily in the belief that the Confederacy had fought to defend states' rights, not slavery. This Big Lie, manifestly untrue but nevertheless told and retold by many defeated Southerners until it acquired the patina of truth, salved their bruised egos, asserted the righteousness of their resistance to "Northern aggression," and prepared for revitalization and rebirth. After Reconstruction, the myth of the Lost Cause exploded in acts of vengeance that overthrew the integrated state governments in the South, enforced Jim Crow laws that kept Negroes "in their place," and propagated the myth on a national scale. The popular success of *The Birth of a Nation* in 1915 sealed the success of a racist caste system that nearly froze White supremacy in place for the next fifty years. The militias, White supremacists, QAnon crazies, and other extreme-right terrorist groups—formerly jealous rivals of each other's success – successfully joined together under the banner of Trump's Big Lie on January 6th. Their unity of purpose makes them much more dangerous in the future than before. Like the South in 1865, these contemporary terrorists know that although they lost the insurrection, this does not mean that they will lose the peace. Many have now tunneled underground to build and propagate a new Lost Cause myth based on Trumpism, awaiting the moment when they can emerge again to fight with altruistic solidarity in the service of authoritarian tribalism.

According to historian David Blight, "The important Lost Causes in history have all been at heart compelling stories about noble defeats that were, with time, forged into political movements of renewal: the French after the Franco-Prussian War of 1870–71 and the profound need for national revanche; Germany after the Great War and its 'stab in the back' theory that led over the 1920s to the rise of the nationalism and racism of the Nazis; and the white South after our Civil War" (Blight 2021: January 10). Blight adds that Trump's insurrection already has many of the necessary elements for a new Lost Cause: "It does not quite have martyrs and a cult of the fallen in which to root its hopes and dreams. But it does have a self-destructive cult leader about to leave power in a defeat that has been transformed into a narrative of betrayal, resistance, and a promise of political revitalization" (2021: January 10). While some political myths are connected to partial truths, others are built entirely on lies and it's already apparent that the stolen election lie of Trumpism may become a gift that will keep on giving. As Blight recognizes, Trumpism knows what it hates, controls major channels of communication, and is firmly rooted institutionally in a defeated and needy Republican Party. Already, says Blight, the Trumpist Big Lie is converting "loss and long-standing grievance into community" and promising to deliver victory "on altars of strife" (January 10). In short, this emergent Lost Cause is poised to deliver shocks of domestic terrorism to American communities and minds on a scale that has not been seen since the Confederate Lost Cause motivated its believers to turn Southern states into racist authoritarian citadels through what they firmly believed were purifying bloodbaths.

In concluding this book, it's important to draw back from this frightening prospect to understand both the coevolutionary stakes of the situation and the possible remedies Americans may turn to in order to counter the continuing downward spiral of social cohesion the nation is facing. I will initiate this turn by comparing Turchin's 2016 *Ages of Discord* with a more recent book that also tracks macro-historical changes over time and proposes possibilities for progressive change to reverse current trends. In *The Upswing: How America Came Together a Century Ago and How We Can Do It Again* (2020), social historian Robert Putnam and his coauthor Shaylyn Garrett celebrate the Progressive Movement that swept US political life from roughly 1900 to 1920. The coauthors begin by sketching their methodology and conclusions, both of which are similar to Turchin's in general ways. In fact, Putnam and Garrett praise *Ages of Discord* in an endnote for also recognizing "the striking concurrence of a multiplicity of factors that followed the same curvilinear course" in twentieth-century US history (Putnam and Garrett 2020: 12). The coauthors assembled an array of statistical evidence to provide a data-based portrait of US history from roughly the end of the Gilded Age in 1895–2020. Specifically, they questioned whether "America has been moving toward greater or lesser economic equality, greater or lesser comity and compromise in politics, greater or lesser cohesion in social life, [and] greater or lesser altruism in cultural values" over the course of these 125 years (Putnam and Garrett 2020: 9).

Like Turchin, the coauthors graphed these indicators by decades and found that economic equality, comity in politics, cohesion in social life, and altruism in cultural values were generally on "the upswing" between 1900 and the mid-1960s. Then, after 1965, the same four indicators underwent declines that gained in steepness the closer they got to the present. Turchin, of course, discerned roughly the same historical turning points separating his integrative and disintegrative trends. Putnam and Garrett named the inverted "U"- shaped curve they derived by synthesizing their four graphs as an "I-we-I" curve (12) because it traces initial progress away from what they call ruinous individualism toward a communitarian pinnacle in the mid-'60s before falling away into chaotic individualism again. Although the coauthors state that "individualists and communitarians can be found on both sides of the political spectrum" (340), it is clear from their indicators that most communitarians live on the liberal side of Tuschman's left-right continuum and many of their individualists are easily located among his conservatives on the right.

Looking more closely at the "pivot" of their graphs in the mid-1960s, Putnam and Garrett note that the 1960s began at a time when the nation undervalued individuality and diversity, especially in the areas of race and gender. By the end of the '60s, however, the nation was celebrating many forms of individuality, especially in the social and cultural spheres. "Nonetheless," they conclude, "the empirical evidence we have gathered in this book also makes it clear that we have paid a high price for the Sixties pivot" (2020: 313). According to their data, this price includes "indefensible economic inequality," "political polarization," "social fragmentation," and "self-centeredness" (313). Although Turchin uses different methods and statistics to look at changes in social cohesion, he would find much to agree within the general conclusions reached by Putnam and Garrett about the 1960s.

Recognizing that their indicators for life in 2020 have taken the nation back to a second Gilded Age at its human worst, the coauthors ask how the progressives of the period managed to transcend or correct the massive problems of the day and move the nation into an era of upswing that lasted roughly seventy years. It's a good question. Like Turchin, Putnam and Garrett recognize that this nation-state has been on a steep decline before and must somehow reverse its downward spiral or fall apart. Both studies work from the evolutionarily sound assumption that the macro-history of organized polities runs in cycles, a truism readily demonstrated by close empirical attention to key indicators.

Putnam and Garrett begin by tracing the "moral awakening" of four progressives (2020: 319–327) who worked in labor reform and government (Frances Perkins), voluntary community services (the initiator of local Rotary Clubs), racial justice politics (Ida B. Wells), and city government (a four-term Mayor of Cleveland). They quote historian Richard Hofstadter on the "moral indignation" of the period (328) and cite similar recent instances, such as the 2018 March for Our Lives to protest gun violence and the Poor People's Campaign, which continues to lead "a national call for moral revival" to combat intergenerational poverty and systemic racism. Other characteristics of effective progressive campaigns included their pragmatic flexibility, their ties to scientific innovation, their ability to work on several issues at once, and the importance of youthful reformers to their success—major attributes they share with moral reformers today. Finally and most importantly for present purposes, progressive politicians could claim substantial moral, social, and economic success by 1920—chief among them, child labor laws, the Pure Food and Drug Act, and constitutional amendments supporting an income tax, the direct election of Senators, and women's right to vote. Overall, Putnam and Garrett find that the progressives reversed the individualistic trends of the Gilded Age and began to emphasize communitarian concerns. In accord with Turchin's political realism and Rutger Bregman's hopeful notion of humankind's "nature," they recognize that norms of social morality always matter in politics.

Despite these general similarities, however, Turchin would likely object that his historical tracking of the progressives reveals that they did not reverse the downward spiral of social cohesion in the US until around 1910 and, equally important, few of their major changes had much to do with communitarian practices and beliefs. Indeed, several of Turchin's trends underlying what he calls "the Progressive Era Trend Reversal" (Turchin 2016: 170) alter little until the 1920s. These include labor-related political violence, general well-being, and suicide rates. Instead, Turchin primarily credits changes in elite values and behaviors for the trend reversal: "The basic argument is that a coalition of elites in America implemented a series of informal reforms, supplemented by a series of formal measures, that were responsible for the reversal of the disintegrative trend" (2016: 171). Some of these measures had to do with progressive legislation, but more of them resulted from the social consolidation of the upper class, a long-term process between 1870 and 1920. This synthesis of norms and behavior featured business mergers after 1900, changing admission policies at elite colleges, and the rising influence of elite economic power in politics. One result of this consolidation was the Immigration Act of 1924. Although massive immigration had created an oversupply of workers after 1900 that kept wages low, the elite knowingly gave up that advantage because of what they believed was a threat to the nation's stability and their growing political power. Many of the newer immigrants from Southern and Eastern Europe had brought anarchism and socialism with them, which—after the Bolshevik Revolution and the bombing of Wall Street in 1920—most businessmen and bankers feared.

Putnam and Garrett admit that compromises on equality and inclusion were an unfortunate result of progressivism, but still find much to praise in progressivism. They admit that many progressives upheld the principles of scientific racism, supported Jim Crow laws and norms throughout the nation, and eventually limited the immigration of Jews, Slavs, and Southern Europeans to the US. The "progressive failure to take full inclusion seriously compromised the integrity of America's 'we' and ultimately sowed the seeds of our subsequent downturn," state the coauthors (Putnam and Garrett 2020: 338). Nonetheless, looking forward historically to New Deal and Great Society programs, they conclude that "in hard measures of economic equality, political comity, social cohesion, and cultural altruism, [the progressives] set in motion genuine upward progress that compounded during the first sixty-five years of the twentieth century" (339).

From my perspective in 2021, Turchin's methodology and reasoning give him a historiographical edge over Putnam and Garrett in this contest between two cliometric arguments about US progressivism. But, as I have already noted, there is substantial overlap between their positions and, to some extent, this is a "glass-half-full or a glass-half-empty" kind of dispute. Neither position rules out the other. Finishing their book before the presidential election, Putnam and Garrett clearly wanted to emphasize that contemporary progressive moral reformers with political savvy had a chance to beat Donald Trump and begin to reverse the acceleration of the slide toward social chaos that the nation has experienced for the last four years. And Biden's success with a moral high ground campaign partly proved them right.

8.6 The Difficult Path Ahead

So my conclusion will borrow from both studies as well as recent history and coevolution to suggest a few lessons for the short term. Like many other nation-states during the last two hundred years, the general coevolutionary tension between tribalism and xenophilia—specifically in the US, between White supremacy and multicultural homophily—will continue to threaten the disintegration of the polity for the foreseeable future. It is a melancholy fact that conflicts centered on racist tribalism underlay the US Civil War and the Reconstruction that followed, pervaded the values and actions of the Southern segregationists who shaped the caste system of Jim Crow and those progressives who passed immigration reform in the 1920s, and that the confluence of these and other tribalist trends led to a high point of US political comity and social cohesion for White citizens only by midcentury. This upswing lasted until 1965 when its indicators began to plummet, due in part to a peaceful struggle for Civil Rights and moderate federal legislation to reverse the nation's systemic racism.

Although the tribalism-xenophilia tension is a universal one that cannot be completely erased, it has been moderated and successfully contained in some nations through the practice of other universal values that affirm human rights, the importance of prestige in relation of authority, and moves toward equalizing social and economic relations. Accordingly, those who instigated and participated in the January 6 insurrection that attempted to eliminate the votes of millions of African Americans in a move to keep President Trump in office need to be brought to justice before political comity can occur. Speaker Nancy Pelosi promises to push for a Truth and Reconciliation Commission to examine the deep causes of the recent insurrection—a second step (after Trump's impeachment trial) along the lines of a similar commission that looked into the causes of the 9/11 catastrophe. This may begin to help the nation grapple with the serious threat of domestic terrorism and point the way toward substantial legislation that ranges from security concerns to gun control and necessary regulations on connective media. Other legislation must be passed, as well, from police reform to economic reparations, to move the nation away from racist traditions toward "a more perfect union."

In the short term, arguments over the charge to this commission (and much else) will no doubt exacerbate the political polarization of the US. Many experts have joined Levitsky and Ziblatt in recognizing polarization as a chief cause of the nation's woes. There are several substantive ways of beginning to address this major problem, which threatens to mushroom into a Lost Cause movement, but no quick fixes. One is reforming the Republican party so that Trumpism dwindles to the point where it is no longer a threat to American democracy. As those running the Lincoln Project understand, appealing to the funders of Trumpism by underlining the public's moral disgust with Trump is one strategy that is already proving effective. Although those in the Lincoln Project hope to return the Republican party to traditional conservatism, progressive Democrats should realize that this is a more likely prospect than the vain hope of converting many present-day Republicans to some version of progressivism. As Tuschman knows, most national populations are naturally divided between liberals and conservatives; shifting political norms over time along the left-right continuum is possible, but wholesale conversion is very unlikely.

The larger question, however, is whether the Republican party—predominately Trumpist, conservative, or something else—can survive at all. Given current demographic trends and voting patterns and assuming that the Democrats will continue to keep voting accessible to nearly all citizens, it is clear that the GOP can probably continue to win elections in rural areas with mostly White voters for another two decades or so, but this is not a sustainable strategy for a national political party. Trumpism will continue because polarization pervades the structures of US politics, likely leaving Republicans with a shrinking minority of voters nationwide and the prospect of permanent minority status. Mitch McConnell knows this and will probably continue to deploy negative partisanship, obstructing Democratic legislation in all ways possible, but this strategy by itself will do little to expand the shrinking Republican base.

Will Wall Street and other Republican funders, however, continue to support such tactics if they come with a Trumpist price tag, especially given the massive problems of the nation and the likelihood of continuing Republican decline? Turchin notes that "some of the reforms adopted during [the progressive era trend reversal], such as the National Labor Relations Act, were highly distasteful to the business and political communities, yet their resistance was overcome" (Turchin 2016: 171). In this regard, it is significant that the National Association of Manufacturers, a reliable Trump ally for the last four years, recently called for his immediate removal from office after the insurrection. The act of these CEOs, aware of the danger to the republic (and also presumably concerned about future profits), makes it likely that other business people and elite groups may also be willing to compromise their short-term economic interests to ensure the long-term stability of the nation. Now may be a good time, especially while the Republican party is in disarray, for Democratic moderates to push those groups who are not so embedded in Heritage Foundation propaganda that they believe they can continue to contain the huge disparities of wealth and power that divide most Americans from the elite without facing dire consequences. Such a strategy would need to be pursued openly without compromising underlying principles to avoid the charge from party progressives that the moderates were abandoning reform to chase after corporate dollars, much as the Clinton Democrats did in the'90s. But the argument for all economic hands-on-deck is a compelling one, especially because the pandemic has so wrecked the economy that even major corporations are finally waking up to the need for systemic change.

In addition, as we know from our deep past and recent history, coalitions among groups with opposing interests have occasionally been one of human nature's ways of effecting lasting change. Some moderate Democrats and Republicans might be able to make common cause with each other, despite polarized political loyalties, for legislation and cooperation across the aisle. While it is too early to know where this might lead, it may pull more Republicans away from far-right extremism and take some of the air out of the Trumpist's Lost Cause and the continuing threat of domestic terrorism.

In any case, the long-term job of persuading Trumpist adherents to abandon the belief that Trump won the election and, consequently, that Biden can never be their legitimate President must begin soon. No one thinks that a Truth and Reconciliation Commission put together by the government will make much of a dent in the beliefs of hard-core Trump supporters, of course. That job is probably best left to family, close friends, liberal churches, psychological services, reformed connective media, and Hollywood. Regarding the dramatic stories we tell each other, a steady diet of *Joker* dysphoria on television and other media would likely be enough to keep both extremes of the left and right locked in their fears and conspiracies through common-enemy identity politics. Thankfully there are at least two other ways of pursuing reform in Hollywood entertainment. First, produce popular books, blogs, and videos that connect the many dots between political extremism and recent Hollywood offerings. Second, patronize, support, and create future Hollywood products that both expose the potential danger and mock the vain foolishness of Trumpist authoritarians. Spike Lee pursued this mode of attack with his comic melodrama *Black Klansman* (2018), which is based on a real incident in 1972 when a Black police officer impersonated a Klansman on the phone and, with his White partner, infiltrated a KKK group in the South that planned to bomb a student Civil Rights group. The two cops stop the attack, but most of the rich comedy of the piece has to do with the stupidity and gullibility of the good ole Klan boys and gals, who finally blow up one of their own cars and three Klan members in an attempt to escape arrest. There are, of course, many other ways of mining similar fictitious situations for comic payoffs, especially if they involve braggart-warrior Proud-Boy types, sleazy-conniving Ted Cruz politicos, profit-hungry advertisers, and other similar low life. As Chaplin understood, standing up to authoritarians can take many forms and one of the most effective to deflate their popularity—and Hitler was popular with many Americans in the late 1930s—may be public mockery.

Another is acts of public mourning. Symbolically, the Biden-Harris inauguration began the evening before the official ceremony, when the President-elect introduced others to help the nation mourn the loss of over 400,000 citizens since the pandemic began. For me, it was a necessary release of tensions and fears I did not even know I had, even though, luckily, I have not yet suffered a family loss from COVID-19. Tears cleared the way for recognizing the deep emotional bonds I share with many of my fellow Americans and prepared me for President Biden's inaugural address the next day, on the 20th. The inauguration took place against the backdrop of the Capitol and facing the national mall, decked out with flags (in lieu of the usual inauguration crowd) and stretching to the Washington Monument with a view of the reflecting pool and Lincoln Memorial behind it. Although this choice had required more national guard troops to secure it, Biden had been right to insist on this majestic and historic setting for the state ritual—only two weeks before the site of the insurrection. With national unity as his primary theme, Biden invited Americans to unite behind the promise of democracy. Although recognizing that such unity may be an "elusive" goal, he urged Americans to "end this uncivil war."

Afterward, thinking of the history and institutional solidity the inauguration had invoked, my thoughts wandered back to pre-Nazi Germany. In 1923, Hitler and his Brown-shirt thugs attempted what became known as "the beer hall putsch" in Munich for its clownish attempt at a coup d'état against the Weimar government. The state put down the insurrection and locked up the perpetrators for a few years—Hitler used his time in jail to write *Mein Kampf* and elaborate his "stab in the back" Lost Cause—but ten years later the Nazis gained legitimate political power and elevated Hitler to the position of Vice-Chancellor of Germany. Many industrialists, Prussian officers, and other elite groups, fearful of the German socialist party, believed they could easily control the political beginner. Soon after he ascended to the Chancellorship with the death of Hindenberg in 1933, however, the Reichstag, the German counterpart to the US Capitol building, mysteriously burned to the ground. Hitler used the excuse of the fire to declare martial law and ended the Weimar Republic a short time later.

I enjoyed visiting the rebuilt Reichstag when I was in Berlin recently and was heartened to hear our tour guide tell this story of the building's history. She also noted that contemporary neo-Nazis would like to burn down the building again, but emphasized that, for Germans, democracy resides in their post-unification constitution and the laws and norms it enables, which includes barring Nazis from political participation. Although the US is unlikely to take this step, the nation should realize that the massive solidity of the Capitol building is no guarantee of US democracy. The Trumpist version of the beer hall putsch of 2021—already much more serious than Hitler's clownish failure—could, with a different, more politically astute neo-Nazi leader, become the Capitol version of the Reichstag fire of 2031 (or much sooner) and end democracy in the US.

References

Balz, D. 2020. Crisis Exposes How U.S. Hollowed Out System. *The Washington Post*, May 17.

Biskind, Peter. 2018. *The Sky Is Falling: How Vampires, Zombies, Androids, and Superheroes Made America Great for Extremism*. New York: The New Press.

Blight, David. 2021. Will the Myth of Trumpism Endure? *The New York Times*, January 10.

Bouie, Jamelle. 2019. The Republican War on Democracy. *The New York Times*, April 28.

Bregman, Rutger. 2019. *Humankind: A Hopeful History*, trans. E. Manton and E. Moore. New York: Little, Brown, and Company.

Cameron, J. 2009. Avatar, U.S. 20th Century Fox.

Cappelen, A., et al. 2020. The Crisis Is Changing How Americans View One Another. *The New York Times*, April 19.

Dean, John W., and Bob Altemeyer. 2020. *Authoritarian Nightmare: Trump and His Followers*. Brooklyn, NY: Melville House.

Devlin, D. 1996. Independence Day. United States: 20th Century Fox.

Dionne, E.J. 2021. The First Step in Moving Forward. *The Washington Post*, January 14.

Jacobson, N., and J. Kilik. 2012. The Hunger Games. United States: Lionsgate Films.

Jones, Douglas A. 2014. *The Captive Stage: Performance and the Proslavery Imagination of the Antebellum North*. Ann Arbor: University of Michigan Press.

Lepore, Jill. 2018. *These Truths: A History of the United States*. New York: W.W. Norton.

Levitsky, Steven, and Daniel Ziblatt. 2018a. *How Democracies Die*. New York: Crown.
———. 2018b. Autocrats Love Emergencies. *The New York Times*, December 23.
Lukianoff, Greg, and Jonathan Haidt. 2018. *The Coddling of the American Mind: How Good Intentions and Bad Ideas are Setting Up a Generation for Failure*. New York: Penguin Press.
Mayer, Jane. 2019. Trump TV. *The New Yorker*, March 11, pp. 40–53.
———. 2020. Back to the Jungle. *The New Yorker*, July 20, pp. 28–39.
Mercieca, Jennifer. 2020. *Demagogue for President: The Rhetorical Genius of Donald Trump*. College Station, TX: Texas A&M Press.
Nolan, C., E. Thomas, and L. Franco. 2005. Batman Begins. United States: Warner Brothers.
———. 2012. The Dark Knight Rises. United States: Warner Brothers.
Osnos, Evan. 2018. Only the Best People. *The New Yorker*, May 21, pp. 56–65.
Poniewozik, James. 2019a. *Audience of One: Donald Trump, Television, and the Fracturing of America*. New York: Liveright Publishing.
———. 2019b. The Real Donald Trump Is a Character on TV. *The New York Times*, September 8.
Putnam, Robert D., and Shaylyn R. Garrett. 2020. *The Upswing: How American Came Together a Century Ago and How We Can Do It Again*. New York: Simon & Schuster.
Reid, Joy-Ann. 2019. *The Man Who Sold America: Trump and the Unraveling of the American Story*. New York: Image Lab Media Group.
Resolution [for Impeachment]. 2021. *The Washington Post*, January 14.
Sandel, M.J. 2020. Are We All in This Together? *The New York Times*, April 19.
Sher, S., R. Hudlin, and P. Savone. 2012. Django Unchained. United States: Columbia Pictures.
Tavernise, S. 2020. Will This Crisis Cement Americans' Lack of Faith in Washington? *The New York Times*, May 24.
Tuschman, Avi. 2013. *Our Political Nature: The Evolutionary Origins of What Divides Us*. Amherst, NY: Prometheus Books.
Thomas, E., C. Roven, and L. Franco. 2005. Batman Begins. United States: Warner Brothers.
Turchin, Peter. 2016. *Ages of Discord: A Structural-Demographic Analysis of American History*. Chaplin, CT: Beresta Books.
Williams, M., and J. Sanchez. 2020. Racism Is a Public Health Crisis. *The Washington Post*, June 5, p. A20.

Index

A
Abramowitz, Alan, 203–205, 228
Aeschylus, 103, 105–107
Ages of Discord, 181, 193, 227, 269, 276
Agriculture, 6, 16–18, 35, 73, 79, 93–95, 124, 250
Alloparenting, 65–67
Alpha male, 6, 12, 16, 67, 68, 72, 73, 80, 86, 96, 98, 103, 109, 110, 131, 142, 173, 189, 231, 232
Altemeyer, Bob, 230, 232, 233, 239, 242
Altruism, 4, 6, 7, 10, 13, 18, 22, 34, 35, 38, 47, 51–54, 57, 58, 66, 68, 87, 103, 108, 183, 250, 267, 268, 276, 279
Anarchism, 278
Anger, 47, 48, 137, 138, 175, 185, 248, 253
Animism, 84

Anti-Communism, 172
Apprentice, The, 229, 230
Archaic states, 37, 76, 79, 96–99
Athens, 101–103, 106, 107, 109, 136
Audiences, 18, 36–40, 42, 48–50, 52, 56, 57, 85, 87, 103–105, 107, 108, 112, 113, 116–118, 120, 123, 124, 132, 134–136, 142, 146, 149, 156, 157, 159–162, 182, 184–187, 189–191, 200–202, 208, 213, 215, 230, 236, 241, 247, 248, 253, 256, 259, 260, 262, 263
Austin, J.L., 84
Austrian empire, 133
Authoritarianism
  authoritarian dominators, 232, 233, 241, 251, 265
  authoritarian followers, 232, 233, 241, 265, 271
  authoritarian hate groups, 237

Authority, 6, 7, 10, 12, 13, 22, 50, 58, 68, 70, 72, 75, 77, 99, 101–103, 106, 110, 112, 114, 119, 120, 128, 130, 132, 138, 151, 153, 154, 171, 180, 183, 184, 187–190, 232, 236, 242, 243, 249, 258, 272, 280
Avatar, 245–248, 250, 252, 254

B

BaYaka Pygmies, 82, 86, 87
Beaumarchais, Pierre-Augustin Caron de, 133, 134, 137
Biden, Joseph, 238, 266, 268–271, 279, 282, 283
Big Lie, 235, 274, 275
Birth of a Nation, The, 275
Biskind, Peter, 240–243, 245, 247, 251–253
Black Lives Matter (BLM), 202, 221–223, 250, 267
Blank slatism, 19
Boehm, Christopher, 5, 6, 82, 86, 250
Bouie, Jamelle, 237, 238
Boyd, Robert, 3–5, 18, 78
Brecht, Bertolt, 153–162
Bregman, Rutger, 249, 250, 278
Bretton Woods Agreement, 170
Buddhism, 109–111, 113, 116
Bushido, 115–124

C

Cameron, James, 245, 246, 256
Cancel culture, 255
Capitalism, 20, 23, 93, 94, 127, 131, 139–142, 152, 157, 159, 169, 178, 195, 201, 203, 214, 217, 218, 246, 250
Capitol building (US), 283, 284

Caste system, 111–115, 122, 124, 127, 140, 214, 248, 275, 279
Catholicism, 128–131, 233
Chikamatsu Monzaemon, 117
China, 16, 18, 95, 97, 99, 100, 118, 120, 122, 123, 128, 150, 151, 172, 264
Christian crusades, 114
Civil war, 100, 114, 129, 147, 151, 152, 172, 204, 209, 211, 269, 270, 274, 275, 279, 283
Climate crisis, 154
Clinton, Bill, 194, 198, 206, 207, 209, 220, 281
Cliodynamics, 11, 15, 16, 18, 21, 22
Coalition politics, 204, 282
Coddling of the American Mind, The, 254, 255
Coevolution, 3–5, 13–16, 20, 30, 50, 59, 63, 65, 69, 75, 203, 249, 279
Cognitive projection, 31
Communism, 153, 169, 171, 172
Competition, 9, 75, 78, 79, 95, 102, 103, 170, 179, 188, 207, 213, 217, 218, 229, 247
Connective media, 216, 217, 220, 221, 230, 251, 254–256, 258, 263, 270–272, 280, 282
Consilience, 24
Constitution (US), 130, 146, 284
Cooperation, 4, 7, 10, 13, 15, 16, 22, 35, 47, 52, 57, 63–66, 68, 71, 78, 99, 132, 163, 213, 214, 247, 250, 252, 256, 262, 282
Cosmopolitanism, 6, 77, 147
COVID-19, 17, 108, 263–269, 283
Cuban Missile Crisis, 172
Cultural studies, 19–21
Culture, 3–6, 10–15, 18–21, 23, 25, 29, 30, 32, 33, 35, 38, 51, 58, 63, 65, 68–71, 74–80, 82, 85,

87, 88, 97, 99, 101, 107, 110, 112, 115, 116, 124, 127, 128, 131, 137, 143, 147, 198, 200, 216, 217, 219, 220, 223, 229, 240, 242, 251, 254, 255, 270
Culture of poverty, 197, 198, 200, 203

D

Da Ponte, Lorenzo, 132
Dark Knight Trilogy,The, 243, 256
Deadpool, 251–255
Demagoguery, 98, 223, 228, 230
Democratic norms, 228, 234, 235, 270
Democratic Party, 177, 203
DeNiro, Robert, 259, 260, 262
Descartes, Rene, 130, 131, 139, 160, 162, 163
Destructive creation, 93, 94, 99, 133, 139
Disgust, 47, 48, 56, 57, 87, 136–138, 147, 159, 161, 174, 181, 182, 186, 260, 267, 280
Django Unchained, 251, 253–255, 258
Doerries, Bryan, 102, 103, 107, 108
Domination, 6–8, 10, 12, 16, 22, 48, 58, 63, 68, 72, 75, 86, 103, 109, 140, 147, 148, 169, 250
Do the Right Thing, 199, 202, 269
Drama, 7, 11, 19, 33, 36, 37, 40, 42, 48, 52, 56, 103, 106, 107, 112, 117, 118, 137, 146, 152, 153, 155, 157, 158, 180–182, 185, 187, 190, 196, 208, 211, 260
Dr. Strangelove, 173–175
Dutch Republic, 139
Dystopia, 243, 247

E

Economics, 14, 20, 160, 243
Egalitarianism, 7, 10, 22, 63, 76, 86, 96, 103, 109, 130, 138, 140, 147, 163, 215, 223, 250, 256
Elite class, 178
Embarrassment, 43, 47, 85, 137
Emotion, 5, 34, 36, 40, 41, 43–48, 50, 52, 56, 57, 66, 68, 69, 75, 88, 98, 112, 134, 136–138, 159, 160, 182, 190, 251, 255, 256, 260, 271
Emotional contagion, 41, 85, 107
Emotion-charged response, 255
Empathy, 30, 40–46, 50, 52, 56, 57, 63, 65, 66, 85, 132, 135, 160, 161, 163, 213, 223, 250, 262, 263, 268, 269
Enaction, 46
Enlightened despotism, 139
Enlightenment, 127, 128, 130, 131, 133, 134, 137, 139, 140, 143, 150, 160–162, 219, 250
Entrainment, 107, 108, 136
Equality, 6–8, 10, 16, 22, 58, 59, 77, 98, 99, 127, 128, 130, 131, 133, 140, 142, 151–153, 163, 176, 203, 213, 221, 249, 276, 278
Ethnocentrism, 58, 71, 77, 80, 190
Evangelical religion, 233, 241
Extremism, 227, 253, 282

F

Facebook, 216, 218–220, 230, 255, 272
Fascism, 174. *See also* Authoritarianism
Federal Bureau of Investigation (FBI), 182, 183, 271, 272
Female solidarity, 86, 138
Floyd, George, 222, 266, 267
Foner, Eric, 213, 214
Fox News, 206, 220, 235–237

French Revolution, The, 127, 128, 143, 227
Full Metal Jacket, 174, 175

G

Gender, 67, 68, 77, 83, 84, 87, 113, 130, 138, 139, 203, 277
Genetics, 3, 6–8, 14, 66, 67, 69, 71–73
Germany, 70, 128, 143, 145–147, 149, 150, 153, 169, 170, 235, 275, 283
Gilded Age (US), 276–278
Gingrich, Newt, 206, 207, 209, 228
Gintis, Herbert, 5, 13–15, 24, 72, 73, 80
Godfather, The, 181, 185, 187–191, 197, 243
God-kings, 96–99, 110
Google, 217–219
Government, 23, 93, 116, 151, 170, 172, 177, 179, 180, 183, 184, 186, 187, 190, 194, 195, 198, 206, 212, 217, 222, 237, 246, 252, 258, 264, 269–271, 273, 275, 277, 282, 283
Great Britain, 17, 141, 144–146, 151, 169
Great Society, 176, 178, 196, 202, 204, 279
Great War, 141, 145, 147, 149, 153, 160, 275
Greece, 107, 124, 150
Grodal, Torbin, 50, 51
Group identification, 33–35, 87, 190
Guilt, 43, 47, 84, 85, 159, 212
Gun control, 270, 280
Gupta empire, 112, 113

H

Health care, 144, 214, 220, 261, 264, 266, 268, 270
Henrich, Joseph, 5, 6, 13, 63, 68, 69, 77–79, 88
Henry V (by Shakespeare), 39, 40, 44, 48–52, 55–58, 212
Hierarchy, 7, 8, 10, 16, 22, 45, 58, 63, 73, 96, 110–115, 119, 128, 130, 138, 140, 195, 213, 217, 221, 232, 240, 243, 250, 258
Hinduism, 111–113
History, 5, 6, 13, 15–18, 20–26, 29, 32, 51, 53, 59, 94, 102, 103, 112, 114, 128, 130, 137, 139, 140, 145, 149, 151, 171, 187, 193, 196, 197, 209, 217–219, 228, 236, 259, 260, 268, 272, 275, 276, 279, 282–284
Hitler, Adolph, 149, 232, 235, 236, 283, 284
Hobbes, Thomas, 249, 258
Hogan, Patrick C., 10, 11, 19, 52, 134
Holocaust, 148, 150, 249, 252
Homo erectus, 12, 64–70, 78
Homo hiedelbergensis, 71
Homo ludens, 15, 80
Homo neanderthalis, 70, 75
Homophily, 6, 7, 10, 22, 279
House of Representatives (US), 206, 237
Hrdy, Sarah, 5, 65
Human nature, 6, 9–12, 19, 31, 39, 44, 130, 198, 240, 245, 248, 249, 258, 282
Human sacrifice, 97, 99
Hunger Games series, *The*, 247, 250, 256
Hunting, 12, 34, 51, 64, 66–71, 77–80, 85, 86, 94

I

Identity fusion, 33, 34, 57, 87, 190
Identity politics, 201–203, 220
  common-enemy, 255, 256, 263, 265, 282
  common-humanity, 255
Ideology, 21, 34, 57–59, 109, 147, 178, 179, 193–195, 217, 232
Impeachment, 206, 207, 238, 272–274, 280
Imperialism, 59, 127, 140–142, 146, 151, 246
Independence Day, 242, 243, 245, 246, 248
India, 10, 96, 109–111, 141, 152
Initiation rites, 35, 84, 85
Insurrection, 271–273, 275, 280, 281, 283
Internet, 36, 203, 207, 214–223, 230, 251, 254–256, 263, 273
Islam, 109, 110, 113, 124

J

Jim Crow era (US), 195, 236
Johnson, Lyndon, 176, 178, 196, 204
Joker, 174, 244, 259–263, 271, 282

K

Kalidasa, 113
King, Martin Luther, Jr., 196, 267
Kolker, Robert, 174, 175
Kubrick, Stanley, 173–176, 245
Kushner, Tony, 209–214

L

Lakoff, George, 9, 11, 119
Language, 7, 10, 20, 22, 31, 53, 57, 58, 74–77, 79–81, 83, 84, 100, 112, 143, 150, 161, 170, 210, 237

Lanman, Jonathan, 31, 33, 35, 89
Leadership, 8, 12, 59, 63, 68, 70, 72, 96, 98, 111, 114, 214, 229, 250
Lee, Spike, 199–203, 208, 282
Left Behind, 231, 241, 242
Lepore, Jill, 205, 207, 219, 220, 231
Levitsky, Steven, 206, 207, 227, 228, 233–236, 280
Libertarianism, 178, 179, 197, 207, 221, 243, 254
Life of Galileo, 127, 153, 154
Lincoln, 209–214, 274
Lost Cause, 274, 275, 280, 282, 283
Low-road capitalism, 201–203
Loyalty, 6, 35, 44, 47, 49, 66, 103, 105, 108, 114–116, 118–122, 145, 149, 153, 158, 189, 203, 204, 228, 233, 255, 274, 282
Lust, 47, 135, 137, 139

M

Machery, Edouard, 9, 10
Macro-history, 277
Manning, Patrick, 22, 178
Markets, 131, 170, 194–196, 217–219, 241, 249
Marriage, 6, 76, 84, 105, 118, 135, 136, 171, 204, 205, 217
Marriage of Figaro, The, 127, 132
Marxism, 157, 160, 162
Marx, Karl, 20
Materialism, 18, 20
Matthew Principle, 95, 96
Mauryan Empire, 100, 109, 110
Mayer, Jane, 178, 180, 195, 236, 265
McAdams, Dan P., 12
McConnell, Mitch, 214, 229, 273, 281
Mega-empire, 98, 100, 109–111, 124
Mental health, 219
Microaggression, 255

Migration, 23, 77–79
Militarism, 124, 169, 171, 174, 253
Monopoly power, 179
Moore, Jason, 94, 139–141, 162, 170, 217
Mozart, Wolfgang A., 132–136, 138
Mueller, Robert, 234
Music, 53, 83, 108, 112, 133, 135, 137, 200, 243, 248, 259

N

Narrative, 7, 31, 40, 45, 50–52, 54, 56, 75, 86, 88, 98, 106, 117, 119, 120, 142, 161, 174, 181, 191, 196, 197, 199, 201, 202, 210, 222, 275
Nationalism, 127, 143, 144, 147, 148, 150, 152, 170, 275
National Security Act, 170
Naturalism, 24–26, 39
Neoliberalism, 23, 177, 178, 191, 193, 194, 217, 224, 264
New Deal, 146, 157, 177, 202, 203, 207, 279
Newspapers, 143, 219, 230. *See also* Print media
Night of the Living Dead, 181, 183, 190, 243
Non-reproductive sexuality, 240
Norms, 5, 7, 10, 14, 34, 35, 38, 43, 45, 63, 71–76, 78–80, 82, 84, 86, 87, 109, 115, 117, 122, 132, 137–139, 159, 183, 185, 190, 204, 213, 214, 216, 217, 222, 223, 228–230, 232, 233, 236, 253, 278, 280, 284

O

Obama, Barak, 205, 206, 211, 214, 220–222, 229, 250, 251
Opera, 131–139

Oresteia, The, 103, 106–108, 136
Orthodoxy, 20, 34, 37, 38, 40, 56, 107, 116, 220, 263
Orthopraxy, 37, 38, 40, 56, 107, 112, 117, 263

P

Pandemic, 17, 23, 263–266, 268–270, 281, 283
Patriarchy, 105, 119, 120, 138, 189, 190, 242, 258
Performances, 7, 13, 15, 21, 24–26, 30, 31, 33, 36–42, 50, 56, 57, 59, 67, 81–85, 93, 101–103, 106–108, 112, 114, 116, 117, 127, 136, 138, 143, 148, 154, 187, 197, 212
    dysphoric elements, 38, 56, 57, 87, 106, 118, 146, 174, 181, 182, 186, 188, 209, 260
    euphoric elements, 38, 40, 56, 57, 87, 106, 110, 124, 174, 183, 191, 209, 260
Persian empire, 100, 103
Piketty, Thomas, 195
Play, 8, 10, 26, 30, 34–40, 42–44, 47, 49, 51–57, 59, 71, 80, 82–84, 86, 87, 102–107, 112, 113, 116–118, 120, 122–124, 128, 131–135, 137, 146, 153–162, 170, 175, 182, 185, 188, 215, 229, 234, 235, 240, 245, 248, 253, 266
Pleistocene Epoch, 7, 13, 15, 48, 63, 64
Poe, Marshall T., 76, 115, 128, 214–217, 223
Police, 37, 145, 157, 172, 176, 177, 182–185, 188, 190, 199–203, 221–223, 245, 258–260, 262, 267, 271, 272, 280, 282

Political polarization, 15, 204, 269, 277, 280
Poverty, 196–202, 221, 277
Prestige, 6–8, 10, 12, 13, 16, 22, 58, 63, 68, 72, 73, 75, 86, 96, 109, 112, 113, 131, 146, 150, 161, 250, 280
Print media, 115
Private property, 95, 143
Progressive era (US), 281
Prosociality, 13, 68, 109
Protestantism, 129, 155
Public sphere, 68–70, 128, 131, 189, 219

Q
QAnon, 271, 275

R
Racial resentment, 205
Racism, 58, 145, 146, 150, 151, 183, 185, 197, 200, 208, 209, 211, 213, 221, 232, 260, 266, 268, 275, 277, 278, 280
Rage, 48, 129, 200, 251–253, 256, 267, 271
Rapture, 241, 242
Reagan, Ronald, 178, 180, 194, 196, 198, 205, 206, 229, 235
Reality shows, 206, 247
Religion, 5, 30, 32, 75, 82, 84, 98, 99, 109–113, 124, 129, 131, 133, 150, 255
Republican Party, 195, 228, 273, 275, 280, 281
Retrodiction, 17, 18, 21, 22
Retrospective prediction, 17. *See also* Retrodiction
Rhetoric, 196, 197, 205, 209, 230, 235, 269
Richerson, Peter J., 3–5, 18, 24, 78

Rituals, 6, 7, 21, 30–36, 38, 40, 42, 56, 57, 63, 67, 68, 70, 71, 79–88, 97, 106, 110, 113, 115, 124, 129, 138, 180, 181, 187, 189, 190, 263
ritual audiences, 36, 37, 56, 87
ritual specialists, 36, 37, 56, 107
Rodgers, Daniel T., 193, 194, 196–198
Roosevelt, Franklin D., 157, 203
Rousseau, Jean-Jacques, 135–137, 139, 249, 258
Russia, 65, 147, 230, 234, 238

S
Samurai, 114–117, 119–122, 124
Sanders, Bernie, 221, 250
Science, 5, 9, 11, 13, 14, 18, 24, 25, 33, 46, 75, 97, 130, 154, 158–162, 180
Secularism, 76
Senate (US), 176, 204, 214, 229, 237, 238, 269, 271, 273, 274
Seshat, 29, 30, 35, 36, 38, 97, 99, 109, 181
Sexuality, 67, 76, 106, 172, 175, 240, 245
Shakespeare, William, 39, 40, 44, 45, 48, 50–53, 57
Shame, 47, 66, 68, 83, 84, 118, 120–122, 137, 159
Smith, Murray, 24, 25
Social cohesion, 15, 21, 33, 34, 37–40, 47, 48, 56, 57, 87, 101, 104, 106, 109, 113, 122, 124, 139, 169, 177, 180, 181, 183, 185, 187, 190, 193, 209, 227, 261, 263, 264, 276–279
Social complexity, 16, 17, 29, 30, 35, 38, 97, 99, 133, 189
Social groups (US)

African Americans, 142, 145, 176, 195, 196, 199, 201, 210, 211, 213, 236, 280
  other American groups, 145
  white Americans, 145, 176, 177, 211, 237, 267
Social identity, 33, 34, 72, 87
Socialism, 145, 278
Sociology, 13–15, 201
Stalin, Joseph, 153, 157, 232
Sterelny, Kim, 4, 5, 71
Stoller, Matt, 179, 180, 209, 218
Structural-demographic theory (SDT), 166
Stuurman, Siep, 130, 131, 150, 151
Superhero films, 251
Supreme Court (US), 176–178, 210, 229, 270
Surveillance capitalism, 217, 218
Sympathy, 41, 43, 45, 47, 66, 121, 135, 181, 187, 222, 250, 268

T
Talk radio, 205, 206
Tarantino, Quentin, 241, 251, 252
Thirteenth Amendment (US), 209, 211, 212, 274
Thirty Years War, 1618–48, 129
Tokugawa shogunate, 115, 116, 128
Tomasello, Michael, 5, 66, 71, 72
Tribalism, 6, 7, 10, 16, 18, 22, 34, 38, 48, 49, 52, 58, 59, 63, 71, 76, 77, 80, 109, 113, 127, 140, 143, 144, 147–149, 151, 172, 213, 231, 232, 240, 250, 258, 267, 275, 279, 280
Triumph of the Will, The, 149, 236

Trump, Donald J., 12, 205, 206, 211, 214, 227–239, 256, 263–269, 271–275, 279–282
Trumpism, 227, 235, 239, 274, 275, 280, 281
Trust, 34, 50, 177, 180, 183, 186, 187, 262, 264, 269, 270, 273
Turchin, Peter, 11, 15–18, 20–22, 24, 26, 29, 35, 38, 57, 93, 95–100, 133, 171, 180, 181, 193, 276–278, 281
  Ages of Discord, 181, 193, 227, 269, 276
  Ultrasociety, 16, 18, 99
Tuschman, Avi, 5–10, 22, 76, 77, 119, 172, 190, 191, 213, 239–241, 258, 277, 280
Twitter, 216, 230, 233, 234, 255, 272

U
Union organizing, 270
United Nations, 150, 242
Universal Declaration of Human Rights, 150
Upswing, The, 276
Urban crime, 243

V
Van Dijck, Jose, 216, 217, 219
Vengeance, 104, 106, 135, 189, 243, 251, 253, 258, 275
Vietnam, 151, 172, 174–177, 179, 189, 203, 243
Vigilantism, 245, 251, 260
Violence, 86, 111, 139, 145, 152, 187, 199, 200, 202, 203, 232, 241, 243, 245, 251, 252,

INDEX 295

256–258, 261, 262, 266, 267, 271, 272, 277, 278
Virtue signaling, 255, 256, 263

W

Wall Street (US), 221, 278, 281
Warfare, 16–18, 34, 53, 77–80, 95, 96, 99, 100, 102, 114, 124, 143, 171, 206, 223, 249
Watergate Babies, 179, 209
Weber, Max, 98
West Wing, The, 207–209
When They See Us, 222, 223
Whitehouse, Harvey, 21, 30, 31, 33–38, 40, 56, 57, 61, 87, 89, 97, 98, 106, 107, 110, 124, 138, 139, 142, 168, 180–183, 187, 190, 241, 261, 263
Wilkerson, Isabel, 140, 221
Williams, Raymond, 18–21
Wilson, E.O., 3, 14, 24, 52, 69
Wilson, Woodrow, 147
Working-class, 37, 144–147, 258, 268
World War II, 169

X

Xenophilia, 6, 7, 10, 22, 34, 58, 59, 63, 76, 109, 127, 151, 163, 213, 250, 256, 279, 280

Y

YouTube, 216, 220, 230, 255

Z

Ziblatt, Daniel, 206, 207, 227, 228, 233–236, 280
Zuboff, Shosana, 217–219

Printed in the United States
by Baker & Taylor Publisher Services